T0336996

PROGRESS IN FUZZY SETS AND SYSTEMS

THEORY AND DECISION LIBRARY

General Editors: W. Leinfellner and G. Eberlein

Series A: Philosophy and Methodology of the Social Sciences
Editors: W. Leinfellner (Technical University of Vienna)
G. Eberlein (Technical University of Munich)

Series B: Mathematical and Statistical Methods
Editor: H. Skala (University of Paderborn)

Series C: Game Theory, Mathematical Programming and Operations Research
Editor: S. H. Tijs (University of Nijmegen)

Series D: System Theory, Knowledge Engineering and Problem Solving
Editor: W. Janko (University of Economics, Vienna)

SERIES D: SYSTEM THEORY, KNOWLEDGE ENGINEERING AND PROBLEM SOLVING

Editor: W. Janko (Vienna)

Volume 5

Scope

This series focuses on the design and description of organisations and systems with application to the social sciences. Formal treatment of the subjects is encouraged. Systems theory, information systems, system analysis, interrelated structures, program systems and expert systems are considered to be a theme within the series. The fundamental basics of such concepts including computational and algorithmic aspects and the investigation of the empirical behaviour of systems and organisations will be an essential part of this library. The study of problems related to the interface of systems and organisations to their environment is supported. Interdisciplinary considerations are welcome. The publication of recent and original results will be favoured.

For a list of titles published in this series, see final page.

PROGRESS IN FUZZY SETS AND SYSTEMS

edited by

WOLFGANG H. JANKO

Department of Applied Computer Science, University of Vienna, Austria

MARC ROUBENS

Faculté Polytechnique de Mons, Belgium

and

H.-J. ZIMMERMANN

Aachen Institute of Technology, F.R.G.

KLUWER ACADEMIC PUBLISHERS

DORDRECHT / BOSTON / LONDON

Library of Congress Cataloging in Publication Data

Progress in fuzzy sets and systems / edited by Wolfgang H. Janko, Marc
Roubens, H.-J. Zimmermann.
p. cm. -- (Theory and decision library, D)
Includes bibliographical references.
ISBN 0-7923-0730-5 (alk. paper)
1. Fuzzy sets. 2. Fuzzy systems. I. Janko, Wolfgang H.
II. Roubens, Marc. III. Zimmermann, H.-J. (Hans-Jürgen), 1934-
IV. Series: Theory and decision library. Series D, System theory,
knowledge engineering, and problem solving.
QA248.P77 1990
511.3'2--dc20 90-4317

ISBN 0-7923-0730-5

Published by Kluwer Academic Publishers,
P.O. Box 17, 3300 AA Dordrecht, The Netherlands.

Kluwer Academic Publishers incorporates
the publishing programmes of
D. Reidel, Martinus Nijhoff, Dr W. Junk and MTP Press.

Sold and distributed in the U.S.A. and Canada
by Kluwer Academic Publishers,
101 Philip Drive, Norwell, MA 02061, U.S.A.

In all other countries, sold and distributed
by Kluwer Academic Publishers Group,
P.O. Box 322, 3300 AH Dordrecht, The Netherlands.

Printed on acid-free paper

Printed in the Netherlands

Contents

List of Authors

Adlassnig, K.-P., Department of Medical Computer Sciences, University of Vienna, Austria.
Bandemer, H., Department of Mathematics, Freiberg Mining Academy, Freiberg, Federal Republic of Germany.
Barreiro, A., Dep. de Electrónica, Fac. de Física, Universidad de Santiago, Santiago de Compostela, Spain.
Bellin, W., Department of Psychology, University of Reading, Reading, England.
Bozicevic, J., Faculty of Technology, University of Zagreb, Zagreb, Yugoslavia.
Campos, L. M., De, Departamento de Ciencias de la Computación e Inteligencia Artificial, Universidad de Granada, Granada, Spain.
Cazenave, M., University of Bordeaux II, Medical Informatic, Bordeaux, France.
Chanas, St., Technical University of Wrocław, Wrocław, Poland.
Couselo, J. M., Dep. de Pdiatría, Hospital General de Galicia, Santiago de Compostela, Spain.
Delgado, A. E., Dep. de Electrónica, Fac. de Física, Universidad de Santiago, Santiago de Compostela, Spain.
Florkiewicz, B., Technical University of Wrocław, Wrocław, Poland.
Fujishiro, I., Department of Information Science, Faculty of Science, University of Tokyo, Japan.
Geyer, A., Department of Operations Research, University of Economics and Business Administration, Vienna, Austria.
Geyer-Schulz, A., Department of Applied Computer Science, University of Economics and Business Administration, Vienna, Austria.
Gonzales Muñoz, A., Dpto. de Ciencias de la Computación e I. A., Universidad de Granada, Granada, Spain.
Hirota, K., Dep. of Instrument & Control Engineering, College of Engineering, Hosei University, Tokyo, Japan.
Kallala, M., Department of Psychology, University of Reading, Reading, England.
Kóczy, L. T., Dept. of Communication Electronics, Technical University of Budapest, Budapest, Hungary.

Kraut, A., Department of Mathematics, Freiberg Mining Academy, Freiberg, Federal Republic of Germany.
Kruse, R., Technical University of Braunschweig, Braunschweig, Federal Republic of Germany.
Kunii, T. L., Department of Information Science, Faculty of Science, University of Tokyo, Tokyo, Japan.
Marín, R., Dep. de Electrónica, Fac. de Física, Universidad de Santiago, Santiago de Compostela, Spain.
Martin-Clouaire, R., Station de Biométrie et Intelligence Artificielle, Castanet-Tolosan Cedex, France.
Meyer, K. D., AEG Aktiengesellschaft, Research Institute Berlin, Berlin, Federal Republic of Germany.
Mira, J., Dep. de Electrónica, Fac. de Física, Universidad de Santiago, Santiago de Compostela, Spain.
Moral, S., Departamento de Ciencias de la Computación e Intelligencia Artificial, Universidad de Granada, Granada, Spain.
Roques, J. C., University of Bordeaux II, Biophysics, Bordeaux, France.
Shirai, Y., Department of Information Science, Faculty of Science, University of Tokyo, Tokyo, Japan.
Spies, M., IBM Scientific Center, Heidelberg, Germany.
Stipanicev, D., Faculty of Electrical Engineering, University of Split, Split, Yugoslavia.
Taudes, A., Department of Applied Computer Science, University of Economics and Business Administration, Vienna, Austria.
Videau, J., University of Bordeaux II, Anatomy, Bordeaux, France.
Vila Miranda, M. A., Dpto. de Ciencias de la Computación, e I. A., Universidad de Granada, Granada, Spain.

Introduction

This volume contains the proceedings of the Second Joint IFSA-EC and EURO-WGFS Workshop on *Progress in Fuzzy Sets in Europe* held on April 6 - 8, 1989 in Vienna, Austria.

The workshop was organized by Prof. Dr. Wolfgang H. Janko from the University of Economics in Vienna under the auspices of IFSA-EC, the European chapter of the International Fuzzy Systems Association, and EURO-WGFS, the working group on Fuzzy Sets of the Association of European Operational Research Societies. The workshop gathered more than 30 participants coming from Western European countries (Austria, Belgium, England, Germany, Finland, France, Hungary, Italy, Scotland and Spain) Eastern European countries (Bulgaria, the German Federal Republic, Hungary and Poland) and non-European countries such as China and Japan.

The 15 selected and refereed papers included in the volume are in principle the author's own versions, with limited editorial changes and small corrections. They are arranged in alphabetical order.

I wish to thank all the contributors for their valuable papers and an outstanding cooperation in the editorial project. I also would like to express my sincere thanks to Professor Dr. H. J. Zimmermann for the cooperation in the refereeing procedure.

Marc Roubens
IFSA-EC and EURO-WGFS
President

UPDATE ON CADIAG-2: A FUZZY MEDICAL EXPERT SYSTEM FOR GENERAL INTERNAL MEDICINE

K.-P. ADLASSNIG
Department of Medical Computer Sciences
(Director: Prof. Dr. G. Grabner)
University of Vienna
Garnisongasse 13, A - 1090 Vienna, Austria

ABSTRACT. A survey on CADIAG-2 a diagnostic consultation system for general internal medicine is presented. The knowledge representation and the inference engine of CADIAG-2 is based on fuzzy set theory and fuzzy logic. CADIAG-2 is integrated into the medical information system of a large hospital. Results obtained by applying the system in the areas of rheumatology, pancreatic diseases, and gall bladder and biliary tract diseases are briefly discussed.

1. Introduction

The central goal of the CADIAG-2[1] project is the development of a medical consultation system for general internal medicine.

Its underlying clinical issues are to assist in the differential diagnostic process:

(a) by indicating all possible diseases which might be the cause of patient's pathological findings, with special emphasis on rare diseases;

(b) by offering further useful examinations to confirm or to exclude gained diagnostic hypotheses or to find stronger support for them; and

(c) by indicating patients' pathological findings not yet accounted for by the expert system's proposed diagnoses.

---

[1] CADIAG stands for *C*omputer-*A*ssisted *DIAG*nosis.

W. H. Janko et al. (eds.), Progress in Fuzzy Sets and Systems, 1–6.

After gaining experience with the medical expert system CADIAG-1 which was formally based on first-order predicate logic and pattern matching (Adlassnig et al. (1985)), a successor system CADIAG-2 was developed and implemented (Adlassnig (1980), Adlassnig (1986)). This system applies fuzzy set theory to model inherent vagueness of medical concepts and fuzzy logic to infer diagnostic conclusions.

At present, CADIAG-2's knowledge base contains disease profiles and complex rules for about 295 diseases, among them 185 rheumatic diseases (69 joint diseases, 12 diseases of the spinal column, 38 diseases of soft tissue and connective tissue system, 45 diseases of cartilage and bone, 21 regional pain syndromes) (Kolarz & Adlassnig (1986)) and 110 gastro-enterological diseases (35 gall bladder and bile duct diseases (Akhavan-Heidari & Adlassnig (1988)), 10 pancreatic diseases (Adlassnig et al. (1984)), 37 colon diseases, 28 disease of the peritoneum).

2. The Medical Consultation System CADIAG-2

2.1. Integration of CADIAG-2 into WAMIS

The CADIAG-2 system is integrated into the medical information system WAMIS[2] of the Vienna General Hospital (Adlassnig et al. (1986)). This integration allows the collection of patient's findings for CADIAG-2 via the routine medical documentation and laboratory system of WAMIS.

Through a data abstraction and aggregation process (Adlassnig (1988)), patient data are made available to the CADIAG-2 system which tries to infer diagnoses from these abstracted findings in a data-driven manner.

In addition, patient data not routinely collected in WAMIS can be added to CADIAG-2 through a man-machine interface which processes medical terms given in natural language. A word segmentation algorithm allows usage of medical synonyms and abbreviations; moreover, it accepts various orthographic variants and takes different medical suffixes into account (Adlassnig & Grabner (1985)).

The CADIAG-2 system was designed in such a way that three modes of application in our hospital are possible:

(a) the screening and monitoring mode applied at a very early stage of the diagnostic process;
(b) the consultation mode applied after complete data collection; and
(c) the textbook mode without connection to the central patient data base.

---

[2] WAMIS is the German acronym for *W*iener *A*llgemeines *M*edizinisches *I*nformations-*S*ystem (Vienna General Medical Information System).

2.2. Knowledge representation and inference engine

CADIAG-2's diagnostic process is based on both stored disease profiles and rules (usually very complex ones such as the ARA criteria for rheumatic diseases (Arnett et al. (1988)).

Two relationships define the association between findings and diseases in these disease profiles:

(a) the necessity of occurrence of a certain finding with a disease (frequency of occurrence degree); and

(b) its sufficiency to infer the disease (strength of confirmation degree).

The same relationships are applicable to define the associations between the antecedents and consequents of rules.

The inference process of CADIAG-2 aims at generating one or more differential diagnoses and—at the same time—at excluding some or all remaining diagnoses. A diagnosis is either established as definitely confirmed or proposed as a diagnostic hypothesis to be confirmed or excluded after additional examinations are performed.

Diagnoses are indicated as definitely confirmed if pathognomonic findings were found in the patient or confirming rules were triggered by patient's findings. Because of the hierarchical relationships among diseases in CADIAG-2, diagnoses at a higher level in the disease hierarchy are confirmed as well if sub-diagnoses are indicated as being confirmed.

Excluded diagnoses are established by either present excluding criteria or absent obligatory criteria. Excluding criteria may be single excluding findings, excluding rules or other, already established diagnoses which exclude other diagnoses. Findings and rule criteria defined to be obligatory present in the patient to establish a certain diagnosis but are definitely absent consequently exclude the respective diagnosis. Definitely excluded disease categories in the disease hierarchy cause also the exclusion of the entire set of the respective sub-diagnoses, if any.

Diagnoses being confirmed and excluded at the same time—which might happen due to contradictory patient data and/or knowledge base errors—are termed diagnostic contradictions. They are displayed separately stating the reason of being established.

Diagnostic hypotheses are generated if a diagnosis is:

(a) neither confirmed, nor excluded, nor a contradictory result; and

(b) the strength of confirmation of at least one present finding, one triggered rule, or one already established sub-diagnosis is equal or higher than a given threshold ϵ $(0 < \epsilon < 1)$.

Since the application of fuzzy set theory allows for mathematical modeling of borderline findings, the degree of presence of a finding (degree of membership in a fuzzy set) is combined with its strength of confirmation. If the resulting value, which is a measure of certainty of the concluded disease, lies between the threshold ϵ and unity (unity means full confirmation), the respective disease has to be taken into consideration as a diagnostic hypothesis.

In addition, diagnostic hypotheses are ranked according to a score of support. This score is calculated on the basis of:

(a) the number of single findings present or present to a certain degree and having a relationship to the disease under consideration;

(b) the degree of presence of these findings; and

(c) the degrees for frequency of occurrence and strength of confirmation between these findings and the respective disease.

Diagnoses which are neither confirmed, nor excluded, nor diagnostic hypotheses, nor contradictory results are put into a category denoted by 'not generated diagnoses'. This allows the physician to obtain a complete survey of all diseases included into CADIAG-2's knowledge base.

In CADIAG-2, two forms of knowledge acquisition have been applied:

(a) acquisition of knowledge from medical experts; and

(b) semiautomatic acquisition of medical knowledge from a patient data base.

Medical experts provide definitional and judgmental knowledge from textbooks and their own practical experience. The estimation of appropriate values for the frequency of occurrence and strength of confirmation degrees is assisted by an automatic procedure which calculates the respective values from stored patient records with known diagnoses (Adlassnig & Kolarz (1986)).

Due to the large number of medical relationships contained in CADIAG-1 and CADIAG-2, intense efforts have been made to verify consistency and completeness of the respective knowledge bases.

For CADIAG-1, a program was developed that verifies the internal consistency of the stored medical knowledge and—in case of inconsistencies—provides the line of reasoning for subsequent correction (Barachini & Adlassnig (1987)). Because of the possible homomorphic mapping of CADIAG-2's finding-to-disease relationships into the finding-to-disease relationship categories of CADIAG-1, this program can partially be applied to CADIAG-2's knowledge base as well (Adlassnig & Kolarz (1986)).

3. Results

At present, extended clinical tests of the accuracy and acceptance of the CADIAG-2 system are being in process (Kolarz & Adlassnig (1986), Adlassnig (1987), Akhavan-Heidari & Adlassnig (1988)).

Results were obtained by applying the system to 544 clinical cases (426 rheumatic cases, 47 pancreatic cases, 71 gallbladder and bile duct cases). Among them, there were 38 multiproblem cases with two discharge diagnoses. For each of these cases, between 200 and 800 findings were available, which were either present, present to a certain degree, or definitely absent. This large number of findings is the result of the complete data collection in the associated medical departments where the tests are being carried out.

The analysis of the CADIAG-2 diagnoses compared with confirmed clinical or, if available, surgery or anatomic-pathological diagnoses yielded an accuracy of about 92%, where the respective evaluation criterion was whether the gold standard diagnosis was either confirmed or among the first three hypotheses in the ranked list of hypotheses. In the multiproblem cases, each discharge diagnosis was evaluated separately.

Very good results could be reached in cases with acute problems and in cases where specific investigations such as X-ray and ultrasonography provided sufficient medical evidence to confirm or to hypothesize a present disease. Unsatisfactory outcome was obtained in some cases with a history of therapy that had let to improved clinical patterns and normalized laboratory test results.

References

Adlassnig, K.-P. (1980) A Fuzzy Logical Model of Computer-Assisted Medical Diagnosis. *Methods of Information in Medicine* **19**, 141–148.

Adlassnig, K.-P. (1986) Fuzzy Set Theory in Medical Diagnosis. *IEEE Transactions on Systems, Man, and Cybernetics* **SMC-16**, 260–265.

Adlassnig, K.-P. (1987) The Application of ROC Curves to the Evaluation of Medical Expert Systems. In *Proc. Medical Informatics Europe '87*, EFMI, Rom, 951–956.

Adlassnig, K.-P. (1988) Uniform Representation of Vagueness and Imprecision in Patient's Medical Findings Using Fuzzy Sets. In *Proc. Cybernetics and Systems '88*, Kluwer Academic Publishers, Dordrecht, 685–692.

Adlassnig, K.-P. & Grabner, H. (1985) Verarbeitung natürlichsprachiger medizinischer Begriffe. In Grabner, G. (Ed.) *WAMIS—Wiener Allgemeines Medizinisches Informations-System*. Springer-Verlag, Berlin, 162–189.

Adlassnig, K.-P. & Kolarz, G. (1986) Representation and Semiautomatic Acquisition of Medical Knowledge in CADIAG-1 and CADIAG-2. *Computers and Biomedical Research* **19**, 63–79.

Adlassnig, K.-P., Kolarz, G., Scheithauer, W., Effenberger, H. & Grabner, G. (1985) CADIAG: Approaches to Computer-Assisted Medical Diagnosis. *Computers in Biology and Medicine* **15**, 315–335.

Adlassnig, K.-P., Kolarz, G., Scheithauer, W. & Grabner, G. (1986) Approach to a Hospital-Based Application of the Medical Expert System CADIAG-2. *Medical Informatics* **11**, 205–223.

Adlassnig, K.-P., Scheithauer, W. & Grabner, G. (1984) CADIAG-2/PANCREAS: An Artificial Intelligence System Based on Fuzzy Set Theory to Diagnose Pancreatic Diseases. In *Proc. Third International Conference on System Science in Health Care*. Springer-Verlag, Berlin, 396–399.

Akhavan-Heidari, M. & Adlassnig, K.-P. (1988) Preliminary Results on CADIAG-2/GALL: A Diagnostic Consultation System for Gallbladder and Biliary Tract Diseases. In *Proc. Medical Informatics Europe '88*, Springer-Verlag, Berlin, 622–666.

Arnett, F. C., Edworthy, St. M., Bloch, D. A., McShane, D. J., Fries, J. F., Cooper, N. S., Healey, L. A., Kaplan, St. R., Liang, M. H., Luthra, H. S., Medsger, Jr., Th. A., Mitchell, D. A., Neustadt, D. A., Pinals, R. S., Schaller, J. G., Sharp, J. T., Wilder, R. L. & Hunder, G. G. (1988) The American Rheumatism Association 1987 Revised Criteria for the Classification of Rheumatoid Arthritis. *Arthritis and Rheumatism* **33**, 315–324.

Barachini, F. & Adlassnig, K.-P. (1987) CONSDED: Medical Knowledge Base Consistency Checking. In *Proc. Medical Informatics Europe '87*, EFMI, Rom, 951–956.

Kolarz, G. & Adlassnig, K.-P. (1986) Problems in Establishing the Medical Expert Systems CADIAG-1 and CADIAG-2 in Rheumatology. *Journal of Medical Systems* **10**, 395–405.

A CASE STUDY ON MODELLING IMPRECISENESS AND VAGUENESS OF OBSERVATIONS TO EVALUATE A FUNCTIONAL RELATIONSHIP

H. BANDEMER and A. KRAUT
Freiberg Mining Academy
Department of Mathematics
P. O. Box 47
9200 Freiberg, GDR

ABSTRACT. In this case study we demonstrate with an example of practical importance how experimental results, given via blurred pictures, can be specified by fuzzy observations and used to evaluate functional relationships. The computation is performed by means of the image processing equipment ROBOTRON A 6471.

1. INTRODUCTION

Usually the results of measurements or observations, shortly data, are recorded as points in an appropriate feature space. To take into account their ever present inaccuracy and uncertainty, they are then considered as realizations of certain random elements. In order to investigate what functional relationship is, approximately, satisfied by the data, we have, as a rule, to specify a family of functions, a so-called setup, from which we want to use a function as an approximation of the unknown functional relationship in question. If it does not make sense to specify such a setup, e.g. in an early stage of investigation, we may use locally fitted regression methods (see Cleveland (1979), Schmerling and Peil (1984)). If we have available an appropriate setup, either a priori from the branch of science or engineering that posed the problem or inspired by the local fit, there is a lot of procedures known in mathematical statistics to solve the problem of approximation (see Bandemer and Näther (1980)). For an interpretation of all the results thus obtained we would need to make some assumptions concerning the random structure of the data, e.g. with respect to the "error distribution" (unless we do feel content simply with making heuristic considerations). These assumptions are, in general, the crucial point of all statistical methods.

An alternative way of dealing with the inaccuracy and uncertainty of the data is opened by their representation as fuzzy sets (Bandemer (1985), (1987), Bandemer and Näther (1988)). In the plane, for example, we may imagine such fuzzy sets as grey-tone pictures, where the degree of blackness of a point corresponds to the grade of possibility that

W. H. Janko et al. (eds.), Progress in Fuzzy Sets and Systems, 7–21.

this point will belong to the observation. While the data, usually, are not recorded in such a form, we must make allowance for the traditional mentality in experimental sciences and suggest methods to specify fuzzy data starting with the usually point-shaped recorded results, taking into account available prior knowledge and reasonable assumptions concerning their inaccuracy and uncertainty (e.g. cp. Bandemer and Roth (1987)). This seems to be a difficult task, which could be compared with that of the specification of statistical assumptions mentioned above. But, there are cases, where e.g. results of measurement or observation are obtained as blurred pictures, the pressing of which to point-shaped results and to statistical assumptions would be quite artificial. Here the situation offers itself to a treatment with tools from fuzzy set theory. Such a case of practical importance will be presented and investigated in this paper.

2. HARDNESS CURVES AT THIN SURFACE LAYERS

One of the most common procedures in testing of material is the measurement of its hardness. Moreover, hardness is frequently a sensitive indicator of material and structural variation when considered in dependence on position in the surface of a specimen or within it.

A usual method to measure ("Vickers"-) hardness consists in the following performance. A regular quadrangular pyramid made of diamond is pressed with a certain fixed power onto the surface of the specimen. When the pyramid is removed the specimen shows a remaining impress of quadrangular pyramidal shape. Hardness is then defined (cp. Siebel (1955)) as the quotient of the pressing power p and the extend of the impress surface, say s,

$$h(p,s) = p/s. \qquad (1)$$

Let d denote the length of the diagonal of the square base of the impress, then we have

$$s = d^2/c_0 \qquad (2)$$

where c_0 depends on the face vertex angle of the pyramid.

In a recent paper (Ohser and Vogt (1982)), the following problem is considered. A rectangular solid specimen was subjected to a hardening treatment onto one of its faces. Then the specimen was cut up orthogonally to this treated face. The inner plane produced in such manner is tested pointwise with respect to hardness, in order to obtain a functional relationship between hardness and the distance to the border, where the hardness treatment was applied, shortly called the depth.

In the paper just mentioned the problem of evaluation was tackled by means of mathematical statistics. The result obtained was an approximating curve composed of cubic splines and reflecting the real conditions of certainty and precision of the measurements to a very low degree, by far too optimistically. Hence we reconsider the problem following up the performance through all stages using tools from fuzzy set theory.

Here, uncertainty and inaccuracy of the empirical results have a lot of sources effecting deviations of these results from some ideal or raising their vagueness or, possibly, both. Whereas the effects of the first kind of sources will be taken into account by our approximation procedure lateron, the effects of the second one will be modelled by introducing fuzzy sets in the next section. As sources essentially of the first kind, we mention deviations from the ideal pyramidal form, local irregularities of material, inclined touching on of the pyramid, mistakes in controlling the pressing power. These sources are, at least by part, under the control of the experimenter and can be influenced and valuated. But the specimen is very small (caused by the intended application of the results) and to assure a high resolving power with respect to position, both, pressing power and impress, must be very small. The observation must be performed by presenting the specimen plane to a microscope and by enlarging the picture by means of an image processing equipment. The result of the observation is presented as the grey-tone picture of the impress on the screen.

The diagonal of the picture is to be measured. Here we find another set of sources for inaccuracy and imprecision, which can not be controlled by the experimenter and which essentially effect the vagueness of the result. We mention the human eye as measuring tool, the light-optical lower bound of resolving power, wavelength of light, adjusting conditions at the microscope, scanning quality of the television equipment. Note, that a sharpening of edges in the grey-tone picture by means of mathematical morphology would raise precision only virtually.

Moreover, the coordinates of the points, where hardness is measured on the specimen, are likewise subject to inaccuracy and uncertainty. The impress has a finite extend and thus measures hardness in a certain neighbourhood of the top of the pyramid. The screen shows always only a part of the plane segment under investigation and turning from segment to segment causes increasing inaccuracy of depths. Hence we will also specify fuzzy sets with respect to measurement of depth.

3.MODELLING THE FUZZY IMPRESS

Every impress caused by the diamond pyramid is represented on the screen of our image processing equipment as a grey-tone picture where the shades reflect the vagueness of the experimental result (see Figure 1). The diagonals of the squares have the same directions as the axes of our coordinate system. Now, each of the pictures is considered a fuzzy set, for which purpose the grey-tone values are normalized to 1. The pictures are denoted by G_j ; $j = 1(1)N$; with the membership function
$g_j \mid \mathbb{R}^2 \rightarrow [0, 1]$.

Figure 1. Vickers-impress in a specimen face

To measure the length of the fuzzy diagonal of G_j we turn to the fuzzy contour C_j of G_j following a suggestion in Bandemer and Kraut (1988). For a closed crisp set G the contour C can be given by

$$C = G \cap \operatorname{clos} G^c \tag{3}$$

where clos G^c means the closure of the complement of G. Hence we may define the fuzzy contour C_j of G_j by

$$C_j = (G_j \cap G_j^{\,c})_{normal} \tag{4}$$

i.e. the renormalized intersection of G_j with its complement. The corresponding membership function is

$$m_{C_j}(x,y) = 2 \min \left\{ g_j(x,y),\ 1-g_j(x,y) \right\} . \tag{5}$$

For the practical performance it is advisable to have the pictures G_j cleaned beforehand from artefacts and basing noise either by morphological transformations such as opening and closing (cp. Serra (1982)) or by simple cutting as given by

$$(g_j(x,y))_{clean} = \begin{cases} 1 & \text{if } 1 - \varepsilon < g_j \\ g_j(x,y) & \text{if } \quad \varepsilon \le g_j \le 1 - \varepsilon \\ 0 & \text{if } \quad g_j < \varepsilon \end{cases} \qquad (6)$$

or the like.

Let y_{Oj} be the coordinate of the horizontal diagonal of C_j. The corresponding membership function $m_{C_j}(x,y_{Oj})$ splits into two parts reflecting the two fuzzy ends of that diagonal. Both these parts can be explained as fuzzy numbers, say M_{Lj} and M_{Rj} . Their difference (cp. Dubois and Prade (1980))

$$D_j = M_{Rj} \ominus M_{Lj} \qquad (7)$$

defines the fuzzy length of the diagonal, the membership function of which is to be calculated by

$$m_{D_j}(d) = \sup_{(x,z):\ d=x+z} \min \left\{ m_{Rj}(x),\ m_{-Lj}(z) \right\} \qquad (8)$$

where m_{Rj} and m_{-Lj} are the membership functions of M_{Rj} and M_{-Lj} , resp. For the case of strictly monotone continuous functions the computation simplifies to

$$m_{D_j}(d) = \alpha \quad \text{for} \quad d = x_{R1} - x_{L2} \quad \text{and} \quad d = x_{R2} - x_{L1} \qquad (9)$$

where $x_{L1}, x_{L2}, x_{R1}, x_{R2}$ are the points of the abscissa with $m_{C_j}(x,y_{Oj})=\alpha$. For further simplification it is advisable to look for (approximate) representations of M_{Rj} and M_{Lj} , which allow for "fast operations" in the sense of Dubois and Prade (1980).

In analogy we can obtain the fuzzy length of the vertical diagonal. If the two diagonals differ in length, this can be caused e.g., by an inclined touching on of the pyramid or by a different hardness depending on the direction, we can extend our model to include this dependence. But, if we are not interested in a further investigation of this phenomenon and if the difference is small, then we can use the arithmetic mean of the two lengths.

When looking forward to the further application of the obtained fuzzy length it is desireable to have an appropriate shape of its membership function, which allows for the use of "fast operations". Oriented at experimental results (cp. Figure 2) it seems justified to choose simple triangular shapes. We specify the shape in dependence on the depth x, where the top of the pyramid touched the plane, i.e. by

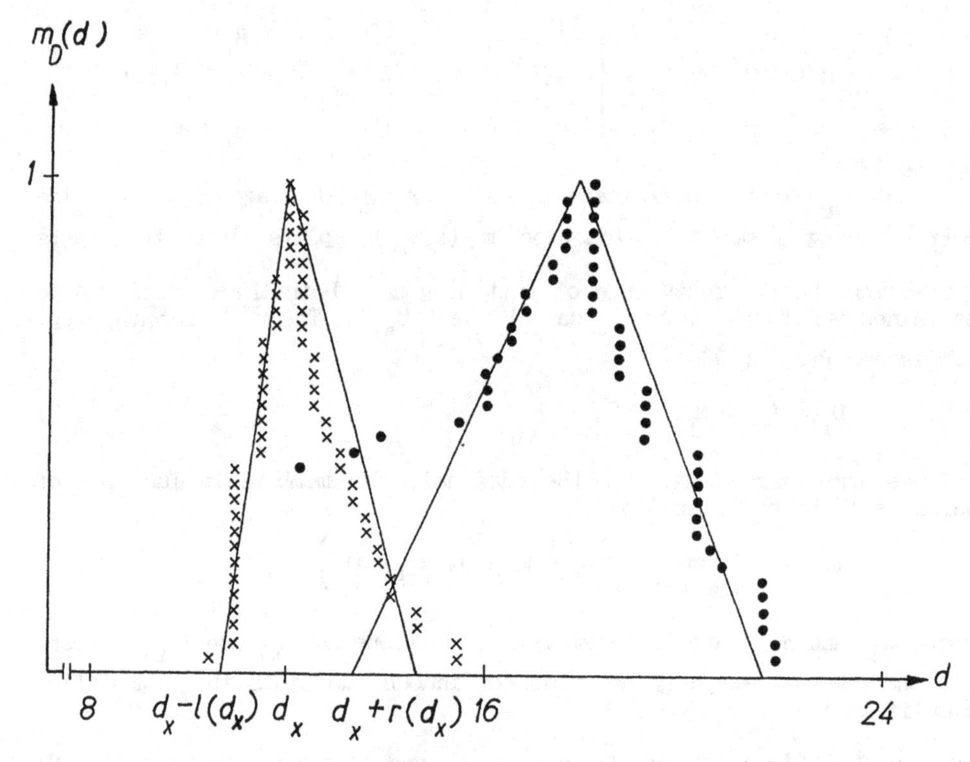

Figure 2. Values of the membership function obtained by measuring (× near the border, ∘ in the kernel) and approximation by hat functions $D_x = [\ l(d_x),\ d_x,\ r(d_x)\ ]$.

$$D_x := [\ l(d_x),\ d_x,\ r(d_x)\ ] \tag{10}$$

where the distances $l(d_x)$ and $r(d_x)$ are the left and the right spread of the support, resp.(cp. Figure 2). The abbreviation D_x is equivalent with the membership function

$$m_{D_x}(d) = \begin{cases} (\ d - d_x + l(d_x))\ /\ l(d_x) & \text{if } d \in [\ d_x - l(d_x),\ d_x\ ] \\ -(\ d - d_x - r(d_x))\ /\ r(d_x) & \text{if } d \in [\ d_x,\ d_x + r(d_x)\ ] \\ 0 & \text{elsewhere.} \end{cases} \tag{11}$$

4. FUZZY HARDNESS AT CRISP DEPTH

We return to section 2 and combine formulae (1) and (2) indicating additionally the dependence on x

$$h(x;\, p,\, s) = c_0\, p \,/\, d_x^{\,2} \tag{12}$$

where c_0, p are crisp constants throughout the investigation and the depth x is assumed crisp for the moment. In the following, up to section 7, we will omit c_0 for convenience.

Now we are in the position to compute <u>fuzzy hardness</u> H(x) at depth x given the fuzzy length D_x of the diagonal using the well-known extension principle (Dubois and Prade (1980))

$$m_{H(x)}(h) = \sup_{z:\ h=p/z^2} m_{D_x}(z) = m_{D_x}(+ \sqrt{p / h}). \tag{13}$$

For our special case (11) we obtain

$$m_{H(x)}(h) = \begin{cases} -(\sqrt{p / h} - d_x - r(d_x)) / r(d_x) \\ \qquad \text{if } h \in [\, p / (d_x + r(d_x))^2,\ p / d_x^{\,2} \,] \\ (\sqrt{p / h} - d_x + l(d_x)) / l(d_x) \\ \qquad \text{if } h \in [\, p / d_x^{\,2},\ p / (d_x - l(d_x))^2 \,] \\ 0 \qquad \text{elsewhere.} \end{cases} \tag{14}$$

Obviously, a crisp length, say $D_x = d_x$, brings back the usual results according to (12).

5. FUZZY HARDNESS AT FUZZY DEPTH

As mentioned in section 2, the depth x must be considered as fuzzy, too. From the manner of ascertaining the depth we assume it also as a fuzzy number with a triangular shaped membership function,

$$X(x) = [\, c(x),\ x,\ c(x) \,] \tag{15}$$

with

$$m_{X(x)}(z) = \begin{cases} (z - x + c(x)) / c(x) & \text{if } z \in [\, x - c(x),\ x \,] \\ -(z - x - c(x)) / c(x) & \text{if } z \in [\, x,\ x + c(x) \,] \\ 0 & \text{elsewhere.} \end{cases} \tag{16}$$

Since there is no reason to assume an interaction between the measurement of hardness and of depth, we can combine the two fuzzy sets to a <u>separable</u> relation

$$H(X) = H(x) \circ X(x) \tag{17}$$

using

$$m_{H(X)}(h,z) = \min \left\{ m_{H(x)}(h),\ m_{X(x)}(z) \right\} . \tag{18}$$

Obviously, the case of crisp depth brings back the usual result according to (12).

For our special membership functions (14) and (16) formula (18) has the following structure

$$m_{H(X)}(h,z) = \begin{cases} -\left(\sqrt{p/h} - d_x - r(d_x)\right) / r(d_x) \\ \quad \text{if } h \in [h_u, h_c] \text{ and } z \in [x - c_1/r(d_x),\ x + c_1/r(d_x)] \\ \left(\sqrt{p/h} - d_x + l(d_x)\right) / l(d_x) \\ \quad \text{if } h \in [h_c, h_o] \text{ and } z \in [x + c_1/l(d_x),\ x - c_1/l(d_x)] \\ (z - x + c(x)) / c(x) \\ \quad \text{if } h \in [p/(d_x + c_2 r(d_x))^2,\ p/(d_x - c_2 l(d_x))^2] \\ \qquad \text{and } z \in [x - c(x),\ x] \\ -(z - x - c(x)) / c(x) \\ \quad \text{if } h \in [p/(d_x - c_2 r(d_x))^2,\ p/(d_x + c_2 l(d_x))^2] \\ \qquad \text{and } z \in [x,\ x + c(x)] \\ 0 \qquad \text{elsewhere} \end{cases} \tag{19}$$

with the abbreviations

$$h_u = p/(d_x + r(d_x))^2;\ h_o = p/(d_x - l(d_x))^2;\ h_c = p/d_x^2$$

and

$$c_1 = \left(\sqrt{p/h} - d_x\right) c(x);\ c_2 = (x - z) / c(x) .$$

Considered as a surface over the (z,h) - co-ordinate-system $m_{H(X)}(h,z)$ looks like a rectangular tent with four curved roof edges from its top.

6. APPROXIMATION BY A FUNCTIONAL RELATIONSHIP

Up to now only one observation, at the fuzzy depth X(x), was considered. To evaluate a functional relationship between hardness and depth we have at first to collect all the fuzzy observations, e.g. by union

$$H = \bigcup_{j=1}^{N} H_j(X) \tag{20}$$

with

$$m_H(h,x) = \max_j m_{H_j(X)}(h,x) \tag{21}$$

as suggested e.g. in Bandemer (1985). To simplify denotation we changed x to x_j and z to x.

For a first impression we consider the maximum trace of $m_H(h,x)$ according to (21), i.e.

$$h_{max} := \left\{ (h, x) \in \mathbb{R}^2 : m_H(h, x) = \max_u m_H(u, x) > 0 \right\} \tag{22}$$

This definition contains the reasonable agreement that intervals, for which there is no information (i.e. $m_H(h,x) \equiv 0$), are excluded from consideration.

This maximum trace is then used to specify a family of functional relationships characterized by only a few parameters, which have a special meaning in the practical context. In our case the following family satisfies these conditions

$$h(x; a, b, v, q) = a + b \exp \left\{ - (x / v)^q / 2 \right\} \tag{23}$$

where a represents the hardness of the kernel of the material (not influenced by the hardening treatment), b the maximal hardness increase, and v and q explain where and how fast hardness decreases with increasing depth (cp. Figure 3).
Given the setup according to (23), for each quadruplet (a, b, v, q) the expected cardinality with respect to the uniform measure over the support S of h_{max} on the x-axis yields a fuzzy set K in the parameter space $(\mathbb{R}^+)^4$:

$$m_K(a, b, v, q) = \int_S m_H(h(x;a,b,v,q),x)\, dx \Big/ \int_S dx \tag{24}$$

with

$$S := \left\{ x \in \mathbb{R}^1 : \max_u m_H(u,x) > 0 \right\} \tag{25}$$

as suggested e.g. in Bandemer (1985).

Since the maximum trace indicates which grade of membership an approximating functional relationship can have at the utmost given the

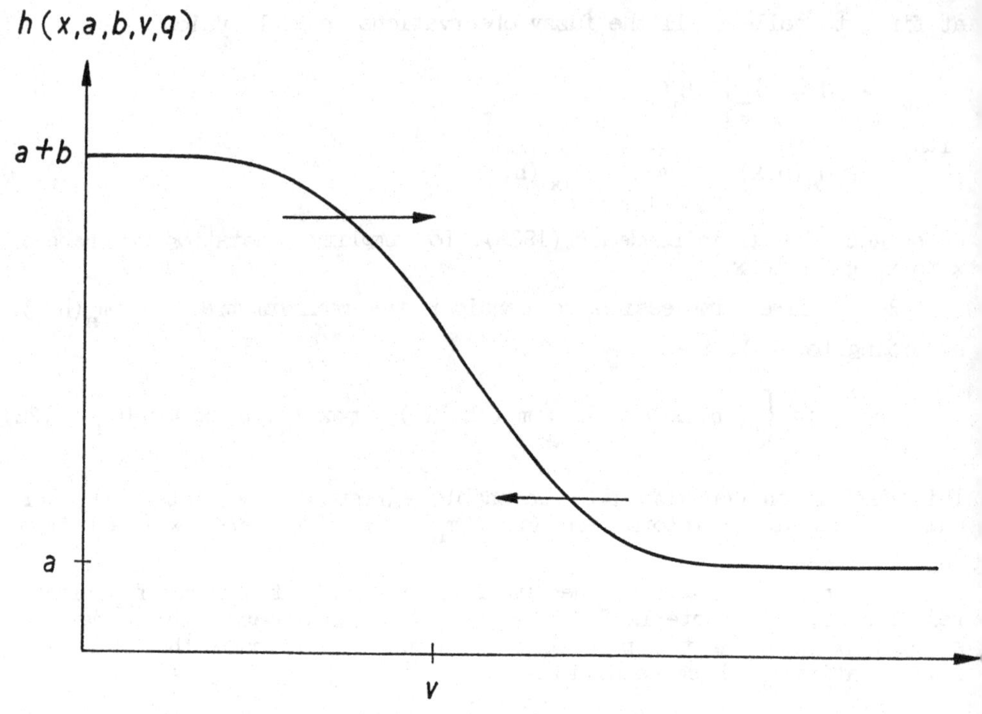

Figure 3. An element of the family of functions according to (23), where the arrows refer to the direction in which the graph will change with increasing q

fuzzy observation, we may take the trace as a standard of comparison and modify (24) according to

$$m_{Krel}(a,b,v,q) = \int_S m_H(\, h(x;a,b,v,q),x)\, dx \;/\; \int_S \max_u m_H(u,x)\, dx \quad (25a)$$

Now, K can be the starting point for further reasoning, e.g. for computing a parameter set K_{max} with maximum membership

$$K_{max} := \left\{ \hat{k} = (\hat{a},\hat{b},\hat{v},\hat{q}) \;:\; m_K(\hat{a},\hat{b},\hat{v},\hat{q}) = \max m_K(a,b,v,q) \right\} . \quad (26)$$

From K_{max} we get most likely graphs

$$\hat{h} = h(\, x;\, \hat{a},\, \hat{b},\, \hat{v},\, \hat{q}\,) \quad (27)$$

for the desired functional relationship (see e.g. Bandemer (1987)).

7. NUMERICAL RESULTS

The method presented and outlined in the preceding sections is now applied to real experimental results.

We started with a cleaning procedure for the impress picture using the transformation

$$(g_j(x,y))_{clean} = \begin{cases} 1 & \text{if } 1-\varepsilon < g_j \\ 2(\ g_j(x,y) + \varepsilon\ ) - 1 & \text{if } 1-2\varepsilon < g_j \leq 1-\varepsilon \\ g_j(x,y) & \text{if } 2\varepsilon < g_j \leq 1-2\varepsilon \\ 2(\ g_j(x,y) - \varepsilon\ ) & \text{if } \varepsilon < g_j \leq 2\varepsilon \\ 0 & \text{if } g_j < \varepsilon \end{cases} \qquad (28)$$

with $\varepsilon \approx 0.2$. This refinement of (6) avoids an abrupt change at ε.

In a preliminary investigation the fuzzy diagonal length were measured in two areas, near the border and far away from it (in the kernel). The membership function values obtained (cp. Figure 2) were fitted by hat functions. The parameters of these functions were averaged for each area separately which resulted in

$$D_{border} = [\ 2.11,\ 12,\ 2.90\ ];\ D_{kernel} = [\ 4.56,\ 17,\ 3.55\ ] \qquad (29)$$

(values given in micrometer). The ranges of l and r near the border were 2.3 and 1.9, resp., in the kernel 2.6 and 4.0, resp.

After a visual inspection we decided that a linear interpolation for l and r in dependence on d_x suffices:

$$l(d_x) = 0.49\ d_x - 3.77;\quad r(d_x) = 0.13\ d_x + 1.34 \qquad (30)$$

Hence for the investigation to follow we had only to measure the values of d_x . In Table 1 we give these values in dependence on the depth. Then $m_{D_x}(d)$ is computed according to (11). The hardness values are obtained via (12) with $p = 0.1$ kp; $c_o = 2 \cos 22^\circ \approx 1.85$, i.e. $c_o p \approx 0.185$.

Inserting (30) and (31) into (14) we get the membership function of fuzzy hardness.

To specify the fuzzy depth we considered the procedure of measuring. The diagonal length d_x was measured automatically within a window of a 160 micrometer width. Then the support of the microscope was shifted program-controlled by 160 micrometer. The maximum error in measuring the depth is 10 micrometer within the first window. The maximum error in shifting amounts to another 10 micrometer. But the shifting errors do not sum up strictly. According to experts´ opinion we

Table 1. Point-shaped results of the hardness measurements d_x in dependence on depth x (both in micrometer)

	x	d_x		x	d_x
1	20	12.92	27	253	16.10
2	21	12.79	28	273	16.23
3	35	12.79	29	277	16.74
4	42	12.41	30	281	17.38
5	46	12.28	31	289	17.12
6	48	12.16	32	308	17.88
7	63	12.41	33	309	17.76
8	68	12.16	34	315	17.62
9	76	12.16	35	324	17.51
10	84	12.54	36	335	17.62
11	94	12.41	37	346	17.62
12	96	12.41	38	350	17.76
13	118	12.54	39	364	17.25
14	120	12.41	40	384	17.51
15	125	13.17	41	393	17.38
16	128	13.43	42	402	17.12
17	154	13.30	43	419	17.76
18	158	13.30	44	435	17.62
19	181	13.43	45	440	17.76
20	183	13.43	46	454	18.02
21	188	13.43	47	471	18.02
22	191	13.43	48	489	17.62
23	215	13.93	49	505	17.51
24	218	14.19	50	516	17.38
25	246	15.59	51	537	17.25
26	249	16.10	52	563	17.51

assumed a maximum error in measuring depth 10 micrometer within the first window, 20 micrometer within the second window and 30 micrometer within each of the following windows.

Hence we specified

$$c(x) = \begin{cases} 10 & \text{if } x \in (\ 0,\ 160\ ] \\ 20 & \text{if } x \in (\ 160,\ 320\ ] \\ 30 & \text{elsewhere.} \end{cases} \tag{32}$$

This was inserted in (16) and with (14) and (16) the membership function of fuzzy hardness at fuzzy depth can be computed according to (19). (See Figure 4)

Now we could look for the desired approximating functional relationship following the line of section 6, using the setup according to (23) and the following approximation of (24), caused by the image

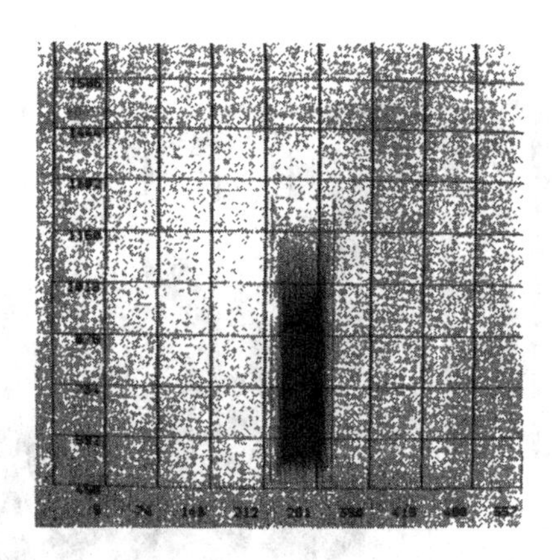

Figure 4. The membership function $m_{H(X)}$ of one observation as presented by the screen.

processing equipment,

$$m_K(a,b,v,q) = \sum_{j=1}^{N(a,b,v,q)} m_H(h_j,x_j) \ / \ N(a,b,v,q) \tag{33}$$

where (h_j, x_j) are the points on the graph of h(x;a,b,v,q) over S and N(a,b,v,q) stands for the number of these points.

By a controlled search we obtained the following results:

Table 2. Results of a controlled search for parameter values of the setup (23) with high membership values with respect to K

a	b	v	q	m_K
610	570	195	5.00	0.711
610	570	197	4.47	0.717
610	570	201	4.34	0.721
610	570	201	4.67	0.721

Figure 5 shows the results as supplied by the image processing equipment, simultaneously the union of the fuzzy observations, the maximum trace, and the "most likely" functional relationship according to the forth row in Table 2. As can be seen from Figure 5, the graph of this functional relationship runs nearly everywhere within the belt of the maximum trace and hence we can assume that m_{Krel} will be nearly one for the corresponding parameters.

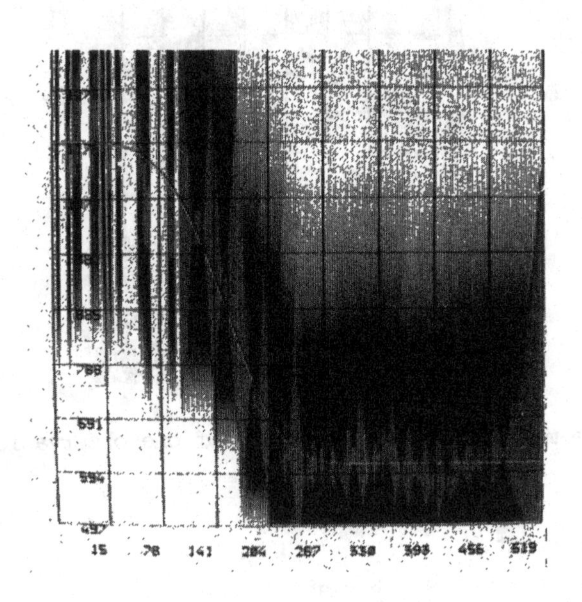

Figure 5. The membership function $m_{H(X)}$, the maximum trace, and the "most likely" functional relationship as presented by the screen.

8. CONCLUDING REMARKS

The advantage of our approach, compared with the usual procedure from mathematical statistics, consists in a more realistic valuation of uncertainty and inaccuracy, which is specified <u>for each single</u> observation and transferred to the parameter space. Moreover, we could introduce a setup making sense in the practical context and being <u>non-linear</u> in its parameters. Finally, we got an impression of the <u>quality of fit</u> embedded in the specified regions of uncertainty and inaccuracy of the given observations. The assumptions concerning shape and extend of the mebership functions can be judged also by the practitioner and the influence of these assumptions can be traced through all computations.

Obviously, our approach is not restricted to the presented example. The increasing employment of image processing equipment and optical sensors offers an extensive field of further applications.

REFERENCES

Bandemer, H. (1985) 'Evaluating explicit functional relationships from fuzzy observations', Fuzzy Sets and Systems 16, 41-52.

Bandemer, H. (1987) 'Fromm fuzzy data to functional relationships',

Mathl. Modelling 9, 419-426.
Bandemer, H. and Kraut, A. (1988) 'On a fuzzy-theory-based computer-aided particle shape description', Fuzzy Sets and Systems 27, 105-113.
Bandemer, H., Kraut, A., and Vogt, F. (1988) 'Evaluation of hardness curves at thin surface layers - A case study on using fuzzy observations - ', Freiberger Forschungsheft D 187, 9-26, Deutscher Verlag für Grundstoffindustrie, Leipzig.
Bandemer, H. and Näther, W. (1980) Theorie und Anwendung der optimalen Versuchsplanung Bd.2, Handbuch zur Anwendung, Akademie-Verlag, Berlin.
Bandemer, H. and Näther, W. (1988) 'Fuzzy projection pursuits', Fuzzy Sets and Systems 27, 141-147.
Bandemer, H. and Roth, K. (1987) 'A method of fuzzy-theory-based computer-aided exploratory data analysis', Biom.J. 29, 497-504.
Cleveland, W. S. (1979) 'Robust locally weighted and smoothing scatterplots', JASA 74, 829-836.
Dubois, D. and Prade, H. (1980) Fuzzy Sets and Systems: Theory and application, Academic Press, New York.
Ohser, J. and Vogt, F. (1982) 'Beitrag zur automatischen Messung und Auswertung von Härteverläufen an dünnen Randschichten', Neue Hütte 11, 428-430.
Schmerling, S. and Peil, J. (1984) Optimal fixing of the bandwidth parameter for the empirical regression', Biom.J. 26, 619-629.
Serra, J. (1982) Image Analysis and Mathematical Morphology, Academic Press, New York.
Siebel, E.(eds.) (1955) Handbuch der Werkstoffprüfung, Vol.2, Springer-Verlag, Berlin-Göttingen-Heidelberg.

****

This article is a shortened version of the paper "Some Applications of Fuzzy Set Theory in Data Analysis", VEB Deutscher Verlag für Grundstoffindustrie, Leipzig 1988.

FUZZY KNOWLEDGE REPRESENTATION FOR LEUKEMIA DIAGNOSIS IN CHILDREN ONCOLOGY

A. Barreiro*, J. Mira*, R. Marín*,
A.E. Delgado* and J.M. Couselo**
*Dep. de Electrónica. Fac. de Física
Universidad de Santiago.
15706 Santiago de Compostela. SPAIN
**Dep. de Pediatría. Hospital General de Galicia
Santiago de Compostela. SPAIN

ABSTRACT. *This article describes the fuzzy, mixed prototype/production-rule methodology employed for knowledge representation and reasoning in ONCOGAL, a system designed to aid in the diagnosis of leukosis. ONCOGAL's diagnostic process involves successive goals set by an automaton structure defined within each prototype. Analysis of the problem of weighting and combining evidence, including the role of expert-engineer communication during the initial installation of static knowledge, has led to the use of a novel method of hypothesis evaluation based on the classification of clinical data in different significance levels referred to by a set of evaluation rules.*

1. Introduction

ONCOGAL is a system designed to aid diagnosis and the monitoring of chemical therapy in paediatric oncology. The scheme used for knowledge representation employs a mixture of prototypes and production rules. .

Certain models of clinical reasoning (Pauker et al., 1976; Elstein et al., 1978) hold that the doctor's diagnostic reasoning is a hypothetico-deductive process: on the basis of the initial clinical data, a number of abductive hypotheses are made in the form of mental prototypes comprising characteristic findings, causal and associative connections, differentiae, etc.; new data whose desirability is deduced from these prototypic hypotheses then bring about convergence on a single well-sustained conclusion. Medical expert systems incorporating prototype-based reasoning of this kind have been described by Pauker et al. (1976), Aikins (1979, 1984), Reggia (1981), Smith and Clayton (1984) and Marín (1987), and experience has shown that for certain fields of medicine this approach is quite adequate. In particular, it may be recommended when existing knowledge of the field in question is strongly structured, even though the structure is largely empirical because of insufficient understanding of the causal mechanisms of the pathologies concerned. A strong knowledge structure facilitates the concentration of knowledge in prototypes, and thus avoids the dispersion of knowledge inherent in the use of production rules for knowledge representation. This difference between the concentration of knowledge in prototypes and its dispersion in production rules means that when the target field is appropriate, basing an expert system on prototypes facilitates both access by the program and interaction with the human expert.

Many prototype-based expert systems evaluate the hypotheses to be chosen from in terms of the coverage of sets of characteristic conditions (Yager, 1985) or by a priori

W. H. Janko et al. (eds.), Progress in Fuzzy Sets and Systems, 22–34.

heuristic means. A weakness of both these methods is that they provide little support for the automatic generation of explanations. In ONCOGAL, two different evaluation schemes have been implemented, one based on a conventional weight combination algorithm, and the other a rule- based method. Both handle uncertainty regarding clinical data and their relevance by means of fuzzy sets (Zadeh 1973, 1979), which have been widely used in rule-based expert systems such as CADIAG-2 (Adlassnig and Kolarz, 1982; Adlassnig et al., 1985a, 1985b).

2. The general structure of the system

The oncologist's knowledge of his field comprises two clearly separate components: knowledge of what clinical variables are relevant, and of how to measure them; and a set of pathology models representing the expectations acquired by personal or collective experience. Both components are represented in ONCOGAL by prototype structures in which each prototype contains both substantive knowledge (the valid values of a clinical parameter, the laboratory data relevant to a particular pathology, etc.) and metadiagnostic knowledge concerning the way in which the substantive data are to be obtained or handled. The metadiagnostic knowledge contained within the prototypes is thus largely responsible for the system's reasoning process, via specific control slots within each prototype and production rules grouped in local knowledge bases referenced by the prototypes. In this way, each pathology prototype is endowed with enough information to enable the objective of the current session for that pathology to be established given the available clinical information; the next step is determined by a transition function on the basis of the results of the current session. In setting up its current objective and determining whether it is fulfilled, each prototype calls upon local rule bases to effect reasoning in areas whose logic is well represented by this kind of structure, such as deduction of the value of one variable from those of others or the generation of hypotheses for differential diagnosis.

As well as substantive and metadiagnostic slots, prototypes also include slots for a variety of non-diagnostic actions that may have to be taken; confirmation of a specific pathology, for example, will trigger the evaluation of factors affecting its prognosis, the generation of reports and the selection of a therapeutic regimen.

Dynamic knowledge, i.e. knowledge concerning particular patients (administrative data, clinical data, previous findings and diagnostic conclusions, etc.), is stored by ONCOGAL in a data base.

At present, ONCOGAL'S knowledge base contains information concerning leukaemias, but the program is in no way limited to this field, and other kinds of cancer will subsequently be provided for. An inbuilt editor allows prototypes and rules to be created, erased and modified by medical staff with only elementary familiarity with the system's syntax.

2.1. CLINICAL VARIABLE PROTOTYPES

A clinical variable is a particular aspect of the patient's condition. Two kinds are distinguished, numerical variables such as foetal_haemoglobin (which takes on values in the range [0,100]), and non-numerical variables such as blood_group (with values A, B or AB) or anaemia (with values YES or NO). Most of the non-numerical clinical variables handled are in fact of this latter, YES/NO kind.

The prototype frame of a clinical variable contains all the information needed in handling it. The generic structure of this kind of prototype comprises the following seven components:

Values specifies the variable's possible values, and is used to check the validity of the data keyed in by the user.

Interaction stores a natural language message used to solicit a value for the variable.

Preference specifies whether the preferred means by which the system obtains the value of the variable is direct (by asking the user, as in the case of fever) or indirect (by deduction from other data, as when the presence or absence of thrombopaenia is deduced from the patient's platelet count).

Clinical action specifies the clinical source of the variable's value: physical examination, blood analysis, reported symptoms, etc. When the system has established that knowledge of the variable will be required during the next phase of the diagnostic process, this component is used to remind the user of the clinical actions that are necessary in order to obtain this knowledge.

Indirect source lists the local production rules to be used for indirect deduction of the value of the variable. Backward chaining is used in their application.

Attributes is a list of subordinate parameters whose knowledge may or must supplement that of the main variable. These subordinate parameters are each endowed with a reduced prototype structure comprising a list of values and an interaction. Fever, for example, has the attributes kind (continuous, intermittent, evening, peaked) and duration_in_days.

Occurrence lists the frequencies with which YES/NO variables take the value YES in relevant pathologies. This information is used in selecting among several possible diagnostic hypotheses.

2.2.PATHOLOGY PROTOTYPES

Pathology prototypes may be related hierarchically as in Figure 1 to reflect relationships of inclusion among the pathologies they represent.

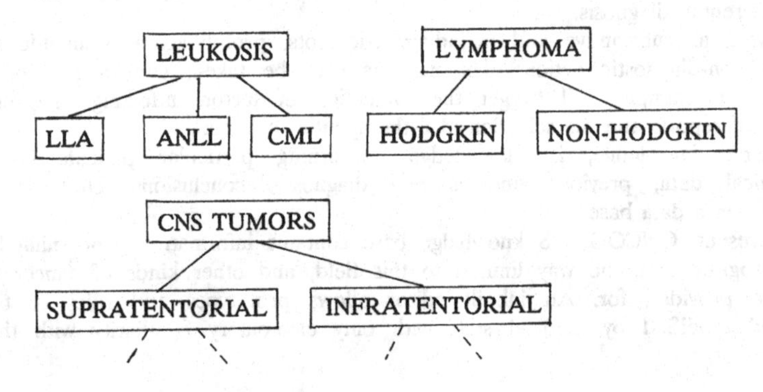

Figure 1. Pathology trees in ONCOGAL.

The generic structure of a pathology prototype comprises the following seven components:

Higher echelon specifies the pathology in which the current pathology is subsumed as a subclass. This allows inheritance of the wider entity's properties.

Lower echelon lists the varieties of the present pathology.

Variables lists the clinical variables that are relevant to the pathology, grouped by the type of clinical procedure by which they are determined. Each variable is listed together with his level of diagnostic significance.

Clinical patterns lists a series of combinations of clinical variables, each of which is of diagnostic significance for the pathology. Not all the patterns associated with a pathology are explicitly listed, since those of the pathology's ancestors in higher echelons are inherited. The combination of variables at the core of the pattern is accompanied by numbers indicating its diagnostic weight and the phase(s) of the diagnostic process to which it is relevant.

Weighting rules lists a series of production rules used in evaluating the relative importance of competing diagnostic hypotheses.

If confirmed specifies the actions to be taken if the pathology is confirmed (warning of possible complications, evaluation of factors affecting prognosis, prescription of therapeutic regimens, etc.).

Control dictates the strategy of the diagnostic process as far as the present pathology is concerned. This strategy is defined in terms of partial goals associated with the states of an automaton structure whose transition from one state to the next depends on the clinical data (Figure 2).

3. The diagnostic process

3.1. CONTROL OF THE DIAGNOSTIC PROCESS

ONCOGAL is inspired by a model of the oncologist's diagnostic reasoning involving two alternating processes, the generation of diagnostic hypotheses on the basis of the available clinical data, and the acquisition of a further batch of data (symptoms, signs, laboratory data, bone marrow data, etc.) in order to test the hypotheses generated in the previous phase of the process. Overall responsibility for the execution of this strategy lies with the main control program, which among other things governs the general manner in which static and dynamic knowledge are accessed, the objectives of the current session established, and diagnostic hypotheses selected. The way in which these general schemes are applied to a particular pathology is nevertheless specified in the control slots of the pathology's prototype. Finally, completion of the diagnostic process triggers the actions specified in the if_confirmed slot of the prototype of the pathology diagnosed.

3.2. LOCAL CONTROL AUTOMATA

Within a given pathology, the alternating process described above means the acquisition of clinical data in successive batches, with considerations of clinical significance, cost, risk and necessity determining which variables are included in each batch. Acquisition of a new batch of data implies a diagnostic objective, the decision whether the data available to date allow the pathology in question to be ruled out, or specifically diagnosed, or whether more data should be sought. Control of this process in ONCOGAL is effected by the automaton structure defined in the pathology's control slot (Figure 2). Each state of the automaton corresponds to a particular state of knowledge as defined by the clinical variables that have been determined, the input to each state is the totality of the relevant clinical data acquired to date, and the inter-state transition function performs the decision as to the next action to be taken. The transition function depends on the value of a degree-of-confirmation function that is cumulative; the degree of confirmation of the pathology is increased by positive clinical results, but is not decreased by negative

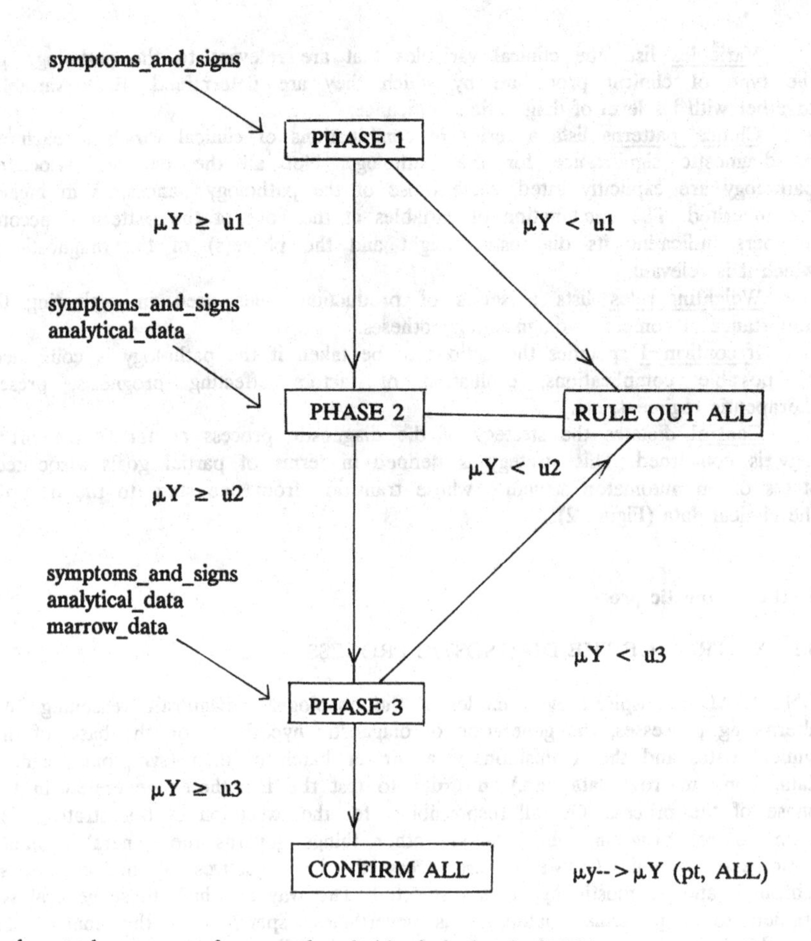

Figure 2. Local control automaton for acute lymphoblastic leukemia (ALL).

results. In keeping with this behaviour, the maintenance of the pathology amongst the current hypotheses requires that the degree of confirmation achieved exceed thresholds that are progressively higher for successive decisions; the pathology is confirmed when the degree of confirmation reaches unity.

3.3. THE DIAGNOSTIC SESSION

The steps involved in each diagnostic session are as follows.

1. The data base is accessed to recover any information concerning the patient in hand. Pathologies that have been ruled out in previous sessions will be ignored in the current session, and within each remaining pathology the diagnostic process is entered at the state determined in the previous session. If there are no data for the patient in the data base, the lowest state of each pathology is entered.

2. The system asks for values of the new clinical variables that are required as input

to the state entered. At this point, the user may also change the values of variables that have already been acquired in previous sessions, which throws the local automaton back to the first state in which these variables are required as input; or feed in the values of variables that have not yet been asked for, which may take the automaton to a higher state.

3. For each pathology in what remains of the pathology tree, the clinical data and clinical patterns relevant to the current state are examined, the degree-of-confirmation is calculated and compared with the threshold for the current state, and the decision is made to rule out the pathology, or to confirm it (in the last state of the automaton) or to ask for more data (in the other states).

4. The patient's subsequent clinical involvement is prepared on the basis of the pathologies that are still in play and their states: analyses required by the next state are listed, or the actions listed in the if_confirmed slot are executed.

4. Fuzzy knowledge representation

The clinician's diagnostic, prognostic and therapeutic judgements and decisions are all based on imperfect knowledge. Uncertainty affects both the values of the clinical variables employed and the empirical relationships among them. The theory of fuzzy sets allows such uncertainty to be handled by computer programs. Fuzzy sets associated with the natural language clinical terms used by the doctor and with natural language expressions of degrees of belief can be fuzzily combined by means of operators for the conjoining of evidence and the global evaluation of all the evidence in favour of a given hypothesis.

4.1. CLINICAL PREDICATES

In order to allow both numerical and nonnumerical data to be handled in the same way, clinical variables are always used in ONCOGAL in clinical predicates of the basic form (variable_name evaluation). When such predicates form part of the system's static knowledge (in the antecedent clause of various kinds of production rule, in the consequent of rules for calculating one clinical variable from others, and in clinical patterns), the evaluation is a value or a range of values, as in (foetal_haemoglobin 15), (foetal_haemoglobin >15), (RH +) or (leukocytosis YES), though the system also allows the latter's being shortened to (leukocytosis). When they occur as part of the dynamic knowledge concerning a particular patient, the evaluation is a list of values together with the degree-of-certainty $\mu \in \{[0,1] \cup \text{"UNKNOWN"}\}$ with which each is attributed to the patient, so that in this case the predicate takes the form

```
(variable_name  (value1 μ(value1)
                ......
                  (valueN μ(valueN)))
```

This form will be interpreted in what follows as a compact way of storing an array of clinical predicates, each with an associated degree of certainty:

```
((variable_name         value1  μ(value1))
 (same_variable_name  value2  μ(value2))
                  ...............
 (same_variable_name  valueN  μ(valueN)))
```

For YES/NO variables, the representation is in fact reduced to

(variable_name YES μ(YES))

since in this case the value NO and its degree of certainty follow immediately. Similarly, since numerical clinical variables are always assumed to be perfectly accurate, only one of their possible values is represented (v1, where $\mu = 1$ for value v1 and $\mu = 0$ otherwise).

The degrees of certainty of non-numerical variables are assigned in three ways, depending on how the values themselves are obtained. a) The values of variables calculated by the system itself from primary source data are associated with fuzzy sets defined on the range of the source data (for example, the value YES of the variable leukocytosis is associated with a fuzzy set defined on the range of values of the numerical variable leukocyte_count); in such cases, the degree of certainty of a particular value for a particular patient is the patient's membership to the corresponding fuzzy set. This is enlarged on in Section 4.3 below. b) The degree of certainty of a YES/NO variable whose value is determined and entered by the user is specified by the user in the form of a natural language label that is associated with a fuzzy set Fi: [0,1] ---> [0,1], and which the system translates into a numerical μ value. c) The degrees of certainty of the values of variables obtained by the system from non-primary source data are calculated using the rules described below in Section 4.3.

When reporting on a variable whose value it has deduced, the system translates a numerical μ value μv as the natural language label corresponding to the fuzzy set Fi for which Fi(μv) is greatest.

4.2. FUZZY COMBINATION OF CLINICAL PREDICATES

In the condition slots of clinical patterns and the antecedents of production rules, clinical predicates are combined by the fuzzy logical operators NOT, AND, OR, MIN and COM. Evaluation of whether a particular patient satisfies such compound conditions consists in using the functions defining these operators to calculate a combined degree of certainty from the degrees of certainty of the individual predicates for that patient. As usual, μ(NOT P) is defined as 1-μ(P), μ(P1 AND ... AND Pn) as min μ(Pi) and μ(P1 OR ... OR Pn) as max μ(Pi). MIN(n1,n2: P1,...,Pn2), which in non-fuzzy terms is true iff at least n1 of the n2 Pi's are true, is defined, following Zadeh, by

$$\mu(\text{MIN}(n1,n2\text{: } P1,...,Pn2)) = \min U(n1,n2\text{: } P1,...,Pn2)$$

where U(n1,n2: P1,...,Pn2) is the set comprising the greatest n1 values among μ(P1), ..., μ(Pn2).

COM(n1,n2: P1,...,Pn2), which in non-fuzzy terms is true iff exactly n1 of the n2 Pi's are true, is defined by

$$\mu(\text{COM}(n1,n2\text{:}P1,...,Pn2)) = \min(\mu(\text{MIN}(n1,n2\text{:}P1,...,Pn2)), 1 - \mu(\text{MIN}(n1+1,n2\text{: } P1,...,Pn2)))$$

Both MIN and COM are familiar logical instruments for the doctor (without, of course, their fuzzy formalization).

4.3. FUZZY INFERENCES AND DEDUCTIONS

Even if the values of clinical variables were known with absolute certainty, uncertainty would still exist as to the reliability of the empirical rules applied to them to obtain diagnostic conclusions. ONCOGAL uses two kinds of inference rule, both of which allow for uncertainty in the transition from premises to conclusions. The values of non-primary clinical variables are deduced by backward chaining.

4.3.1. *Deduction of non-numerical data from numerical data.* Basic numerical data are assumed to be known for certain; uncertainty appears in the deductive process when non-numerical data are deduced from the numerical facts. An example is the deduction of the existence of thrombopaenia from a platelet count; the platelet count. is assumed to be reliable, but whether a given count amounts to thrombopaenia depends in an ill-defined fashion on the characteristics of the patient. ONCOGAL handles this kind of situation by associating the values of the non-numerical variable with fuzzy sets. Thrombopaenia, for example, may be defined by the fuzzy set which membership function is 1 - S(x; 20,55,90), where x is platelet concentration in thousands per cubic millimetre and S is Zadeh's standard function (Zadeh 1979), defined by

$$S(x; a, b, g) = \begin{cases} 0 & x \leq a \\ 2[(x-a)/(g-a)]^2 & a < x \leq b \\ 1 - 2[(x-g)/(g-a)]^2 & b < x \leq g \\ 1 & x > g \end{cases}$$

Such a smooth function is generally not necessary, however, and it is sufficient to use a linear function defined by a cutoff value xc = 20 and a slope s = -1/70.

Any available knowledge as to how the patient's characteristics affect the relationship between the numerical and non-numerical variables can be included in the above deductive mechanism by varying the parameters defining the fuzzy sets (a, b and g, or xc and s). In particular, in paediatric oncology it is common for age and sex to have a bearing on the interpretation of data. The thrombopaenia functions shown above, for example, are valid for patients older than 16 years.

4.3.2. *Deduction from non-numerical data.* ONCOGAL uses production rules of the conventional form RIk = "IF <condition> THEN <clinical_predicate>" whenever the deduction of a clinical predicate can be concisely expressed in this way, as for the rule "if the erythrocyte count is low or the haemoglobin level is low, then the patient has anaemia". The reliability of each such rule is expressed by an associated degree of certainty μRIk, and the degree of certainty of the clinical predicate deduced is calculated as min (μRIk, μ(condition)), where μ(condition) is obtained as in Section 4.2.

All the rules concluding values of a given clinical variable are listed in the indirect_source slot of that variable's prototype frame. When more than one rule affects the same clinical predicate, the joint degree of certainty of the latter is calculated as the greatest of the degrees of certainty derived from the individual rules.

4.4. FREQUENCIES OF OCCURRENCE AND CONFIRMATION

So far, we have mainly dealt in Section 4 with the fuzzy relationship between the patient and the values of clinical variables. Except when patognomic variables are involved, uncertainty also exists regarding the relationship between pathologies and clinical variables, whether single or combined in clinical patterns. In the case of single variables, biomedical statistics usually provide a reliable measure of such uncertainty, the frequency with which each value occurs within the population of patients suffering the pathology in question

(Adlassnig and Kolarz, 1982). This is the frequency stored as a percentage associated with the pathology in the occurrence slot of the variable's prototype frame. We shall see below that during the evaluation of diagnostic hypotheses it is used basically in a negative way: non-satisfaction of clinical predicates that are typical of the pathology confirms that the pathology can be ruled out.

There are very few data available concerning the frequency with which clinical patterns combining several variables occur within pathology groups, and even fewer concerning the frequency with which a given pathology is confirmed among the group of patients presenting a given clinical pattern, which would be the direct measure of the degree to which the clinical pattern supports diagnosis of the pathology. In order to endow the expert system with a measure of this support during its creation, it is at present usually necessary to resort to the human expert's subjective estimate of the frequency with which the pathology is confirmed, given the clinical pattern. This estimate, expressed as a percentage, is the diagnostic weight stored in the confirmed slot of the clinical pattern.

During the operation of the system, both the above percentages are converted by the function S(x;1,50,99) to measures in the range [0,1] to which fuzzy logical operations can be applied: the degree-of-occurrence, μO, and the diagnostic weight μC.

TABLE 1. Levels of diagnosons for ALL

LEVEL 0
Pallor
Anorexia
Fatigue
Stomach ache
Aching bones
Aching joints
Loss of weight
Low coagulation factor level
Consumption of coagulation factors
LEVEL 1
Fever
Haemorrhage
Anaemia
Thrombopaenia
Leukocytosis
Leukopaenia
LEVEL 2
Hepatomegalia
Splenomegalia
Adenopathy
LEVEL 3
Lymphoblasts in peripheral blood
Lymphoblasts in bone marrow over 25%

5. Evaluation of diagnostic hypotheses

The basic clinical problem in the diagnosis of many pathologies is that determination of the definitive pathognomic data (in the case of leukosis, the presence of lymphoblasts in bone marrow) is either expensive or puts the patient at risk. As a result, as much diagnostic information as possible must be squeezed from more readily available data so as to avoid unnecessary costs in money or discomfort. The diagnosis of acute lymphoblastic leukaemia, for example, involves in practice not only the pathognomic variables lymphoblasts_in_peripheral_blood and lymphoblasts_in_bone_marrow_over_25%, but also some 18 other YES/NO variables representing different possible features of the disease (Table 1). In what follows we shall term such features "diagnosons". Sections 5.1 and 5.2 below describe two different strategies by which the values of diagnosons can be used to evaluate competing diagnostic hypotheses. In order to allow their comparison, both have been implemented in ONCOGAL.

5.1. EVALUATION OF DIAGNOSTIC HYPOTHESES USING CLINICAL PATTERNS

Since not all diagnosons are invariably present in a given case, since not all can be considered in the early stages of diagnosis, and since many, if not most, may be individually relevant to other pathologies, the practising clinician may work by deciding whether the available information for a given patient fits any of a number of clinical patterns that the suspected pathologies are known to induce among the diagnosons relevant to the current stage of the diagnostic process. Since a complete set of clinical patterns covering all possible eventualities would be prohibitively large, the problem is to find a sufficiently small set that nevertheless reduces errors to within the limits of current medical knowledge. Essentially, this is attempted by grouping together patterns which have similar diagnostic weights for the pathologies considered and whose component diagnosons are mostly the same. In particular, patterns with small diagnostic weight may be eliminated altogether. Other variations of this reduction strategy can be formalized as follows by considering the clinical patterns as initially expressed as strings of diagnosons linked by AND and NOT operators.

a) If a diagnoson d1 is considered to have but slight importance given d2 and d3, then the diagnostic weight of pattern 1 = (d1 AND d2 AND d3) will be close to that of pattern 2 = (NOT d1 AND d2 AND d3). Patterns 1 and 2 may therefore be eliminated and replaced by pattern 3 = (d2 AND d3).

b) Patterns with similar weights may be combined by using additional logical operators such as OR, MIN or COM. Thus (d1 AND d2) and (d1 AND d3) can be merged as (d1 AND (d2 OR d3)).

Once the set of clinical patterns has been established, the evidence in favour of the various pathologies in a given case can be quantified using their diagnostic weights. In ONCOGAL this is done by adapting the conventional use of fuzzy weighting functions (Adlassnig and Kolarz 1982) to ONCOGAL's prototype-based knowledge representation scheme. Specifically, for each of the competing diagnostic hypotheses (pathologies) Dj that occupy terminal positions in the current pathology tree (Figure 1), a degree-of-confirmation μY and a degreeof-exclusion μN are calculated using the formulae

$$\mu Y(pt,Dj) = \max_{i}(\min(\mu p(pt,\text{pattern } i),\mu c(\text{pattern } i,Dj)))$$

$$\mu N(pt,Dj) = \max_{i}(\min(1-\mu p(pt,\text{diagnoson } i),\mu o(\text{diagnoson } i,Dj)))$$

where "pt" indicates the patient, μp is the degree of certainty of the patient's fitting a given diagnoson or pattern (Sections 4.1-4.3 above), and μo and μc are respectively the degree of occurrence of a diagnoson in pathology Dj and the diagnostic weight of a pattern for pathology Dj (Section 4.4). For a non-terminal pathology, μY is the greatest of the μY of its subtypes, and μN the least of the μN of its subtypes. A pathology is deleted from the tree for the current patient if $\mu N = 1$ or if μY is less than the threshold for the current phase of the diagnostic process (Section 3.2); regarded as confirmed if $\mu Y = 1$; and maintained as an open hypothesis if μY is less than 1 but greater than or equal to the current threshold. Confirmed hypotheses are announced as such; in their absence, the open hypothesis with the greatest μY is suggested as being currently the best.

5.2. EVALUATION OF HYPOTHESES USING EVIDENCE WEIGHTING RULES

The main drawback of the clinical pattern approach described above is the danger of oversimplification in reducing the set of patterns to a manageable size. The replacement of several patterns with different diagnostic weights by a single wider pattern limits the sensitivity of the method, especially when the patterns' fuzzy nature is not taken into account during the reduction process, and great demands are put upon the medical staff collaborating in the installation of the knowledge base. Misunderstanding between the doctor and the knowledge engineer may lead, for example, to "d1 and d2" being installed as (d1 AND d2) instead of (d1 AND d2 AND NOT d3) when the NOT d3 part of the latter is not insignificant. As a result of these difficulties, important patterns may be left out of the knowledge base and, contrariwise, sub-patterns ORed into larger patterns may be afforded undue weight (in which case the max-min rule for μY may lead to undue weighting of one of the competing hypotheses). It is in fact found that during the intermediate stages of diagnosis, when pathognomic data are still not available, the clinical pattern method of Section 5.1 often announces a "best hypothesis" that differs significantly from the human expert's opinion; as a result, the system may give priority to obtaining further data that are irrelevant to the pathology regarded as most likely by the clinician.

The problems associated with the use of clinical patterns may be obviated by using a hypothesis evaluation function defined, like μN above, in terms of individual diagnosons rather than patterns, but which takes into account the generally non-linear importance of combinations of diagnosons. Ad hoc functions of this kind have been used in the expert systems PIP (Pauker et al., 1976), CENTAUR (Aikins 1979, 1984), CAEMF (Marín 1987) and CADIAG-2 (Adlassnig et al., 1985). CADIAG-2, for example, uses the evaluation function

$$PN = \sum_{i} [a*\min(\mu p(pt,di),\mu o(di,Dj)) + b*\min(\mu p(pt,di),\mu c(di,Dj))]$$

where μc is now the diagnostic weight of the "pattern" consisting of a single diagnoson, and a and b can be varied (subject to $a + b = 1$) so as to adjust the relative importance of degrees of occurrence and diagnostic weights; the best hypothesis is chosen to be that with the greatest value of PN. Unfortunately, the use of such functions makes it virtually impossible for the user to find out just why (in terms of diagnosons satisfied) one hypothesis is preferred to others. This both hampers the development of the system and leads to users treating the system with extreme wariness.

The alternative method used in ONCOGAL is based upon the grouping of diagnosons, not in clinical patterns, but in levels of diagnostic significance (Table 1) referred to by a set of evaluation rules of the form

```
REk:IF   qi1(level1)
         AND qi2(level2)
```

....

AND qiN(levelN)
THEN accept hypothesis Dj
with confidence μREk

where each qi is a fuzzy quantifier and μREk is the support for Dj provided by Ek, the set of diagnosons considered in the rule. A typical evaluation rule, in natural language form, is

REk: IF almost all Level O diagnosons are present
AND some Level 1 diagnosons are present
THEN diagnose acute lymphoblastic leukaemia with confidence μREk

The fuzzy quantifiers qi are defined on [0,1], the range of the degree of certainty of the patient's globally satisfying the conditions of any given level, which is defined as the average of the degrees of certainty with which he or she satisfies the individual diagnosons included in that level. The degree of certainty of the patient's complying with the antecedent as a whole, μp(pt,Ek), is calculated from those of the clauses for each level using standard fuzzy logic. The final support for Dj is defined as

$$\mu Y(pt,Dj) = \max_{k}(\min(\mu p(pt,Ek),\mu REk))$$

With this method, the system's choice of a particular hypothesis as the best can readily be explained to the user by displaying the natural language form of the evaluation rules employed together with the list of diagnosons present or absent and their significance levels.

6. Conclusion

The expert system ONCOGAL, which is currently oriented towards the diagnosis of leukosis, employs fuzzy knowledge representation by prototypes. Sequences of diagnostic objectives defined for each pathology included in the system allow considerations of cost, risk and necessity to be taken into account during the diagnostic process when requesting further clinical data. The storage of the necessary information in prototype frames for each pathology and clinical variable facilitates expansion of the knowledge base.

Analysis of the problems met by conventional methods for evaluating competing diagnostic hypotheses, including their sensitivity to errors during the installation of static knowledge in collaboration with human experts, has led to the design of an alternative method in which a set of evaluation rules enables global consideration of relevant clinical features. In the future development of the system, it is intended that hypothesis evaluation should also take into account non-statistical factors such as causal relationships among clinical features.

7. References

1. Adlassnig K.P. and Kolarz G. (1982) "CADIAG-2: Computer assisted medical diagnosis using fuzzy subsets". Approximate Reasoning in Decision Analysis. North-Holland Pub. Co., 219-247.

2. Adlassnig K.P., Kolarz G., Scheithauer W., Effenberger H. and Grabner G. (1985) "CADIAG: Approaches to computer-assisted medical diagnosis". Comput. Biol. Med., 15, 315-335.
3. Adlassnig K.P., Kolarz G. and Scheithauer W. (1985) "Present state of the medical expert system CADIAG2". Meth. Inform. Med., 24, 13-20.
4. Aikins J. (1979) "Prototypes and production rules: An approach to knowledge representation for hypothesis formation". Proc. of IJCAI-79.
5. Aikins J. (1984) "A representation scheme using both frames and rules". In Rule-Based Expert Systems, B. Buchanan and E. Shortliffe (eds.). Addison-Wesley.
6. Elstein A., Shulman L. and Sprafka S. (1978) "Medical problem solving". Harvard University Press.
7. Marín R. (1978) "Un sistema experto para el diagnóstico y tratamiento anteparto del estado materno-fetal". Doctoral Dissertation. Universidad de Santiago.
8. Pauker S., Gorry G., Kassirer J. and Schwartz W. (1976) "Towards the simulation of clinical cognition: taking a present illness by computer". Am. J. Med., 60, 981-986.
9. Reggia J. (1981) "Computer assisted medical decision making: a critical review". Annals of Biomed. Eng., 9, 605-619.
10. Smith D. and Clayton J. (1984) "Another look at frames in rule-based expert systems". B. Buchanan and E. Shortliffe (eds.). Addison-Wesley.
11. Zadeh L.A. (1973) "Outline of a new approach to the analysis of complex systems and decision processes". IEEE Trans. on Systems, Man and Cyb., SMC-3, No. 1, 28-44.
12. Zadeh L.A. (1979) " A theory of approximate reasoning". In Machine Intelligence, 9, J.E. Hayes, D. Michie and L.I. Kulich (eds.). Wiley, New York. 149-194.
13. Yager R.R. (1985) "Explanatory models in expert systems". Int. J. Man-Machine Studies, 23, 539-549.

TOWARDS A FUZZY MODELING OF THE FUNCTIONAL ANTHROPOMETRY

M.CAZENAVE
University Bordeaux II
Medical Informatic
33000 BORDEAUX
FRANCE

J.C.ROQUES
University Bordeaux II
Biophysic
33000 BORDEAUX
FRANCE

J.VIDEAU
University of Bordeaux II
Anatomy
33000 BORDEAUX
FRANCE

ABSTRACT. This paper is the starting point of the comparison between the most meaningful anatomic parameters of two groups:sportsmen and nosportsmen.The anatomic parameters are selected as interdependent.The first step is to construct the membership function of each anatomic parameter viewed as a fuzzy subset.We emphasize with details how to choose some criteria optimizing the membership function.The second step is the approach of the group's membership function.We specify the algorithm we need in this view.Then we present first results from the group's membership functions and discuss the future prospects.

1.INTRODUCTION.

Among the problems concerning the morphologists (ANTHONY, 1903),there is one which arrests their attention:the classification of the human types.The physical characters were at the starting point of the first classifications.Later with a view to the personality notion came the physiological so as psychic characters.

The discriminatory characters between the humans take place in two families:

-hereditary characters as coloured people;they are in the field of the Anthropology.

- acquired characters as stature of weight. The biotypology makes a special study of this field (BARBARA,1929).

The morphogramm is th whole of the anatomical measure-

W. H. Janko et al. (eds.), Progress in Fuzzy Sets and Systems, 35–46.

ments,which are necessary for defining any human group from the biotypological point of view.

BARROIS (1979) describes and compares the anatomical me asurements of 436 male humans,aged from 17 to 25 years. 143 of them are students of the military health corps School. 293 are students of the regional Center of physical and sports T-raining.The sportsman is the one practising regularly the com petition;he began the physical training very soon (between 6 and 10 years).100 students of the military health corps S-chool are viewed as nosportsmen because they practise the S port only in their leisure hours.336 students are viewed as sportsmen.BARROIS (1979) studies the means,standard-deviati ons and statistical correlations of twenty anatomical para-meters and shows that only ten of them are worth keeping be cause the correlations between these parameters are from 0,3 to 0,7 and make them indissociable in the defining processes of the group's morphological type.These anatomical parame-ters are:

- -the weight,
- -the full height,
- -the seating height,
- -the biacromial distance,
- -the rib cage's transverse diameter at the level of the sternum-xiphoid articulation,
- -the rib cage's depth,
- -the distance between the summits of the iliac c-rests,
- -the distance between the trochanters,
- -the thoracic member's length,
- -the pelvic member's length.

These parameters are represented on the figure 1. With the angle between the pubis and the iliac antero-superior tube-rosities,they allow us to exhibit the human's morphological polygon (figure 2).

The means and standard deviations of these parameters introduce the anatomical morphograms of each group (OLIVIER and PINEAU (1961),BARROIS (1979),BONJEAN and ROQUES (1983)). The statistical comparisons of the parameters between any-one group and each other complete these descriptions.

We intend in this paper to prolong this statistical a-nalysis by fuzzifying each anatomic parameter.In such a pro cess the anatomical parameter becomes a fuzzy subset with a membership function.The first step is to construct this mem ship function.

In the second step we have to construct the algorithm, valuing the membership of each group (sportsmen,nosportsmen and general population defined as both of the precedings), from the membership functions of all the anatomical parame-ters viewed as fuzzy subsets.These parameters are mutually dependent in the sense of the statistical correlations.

The presentation of some results and the discussion of the choices we needed in view of the membership functions

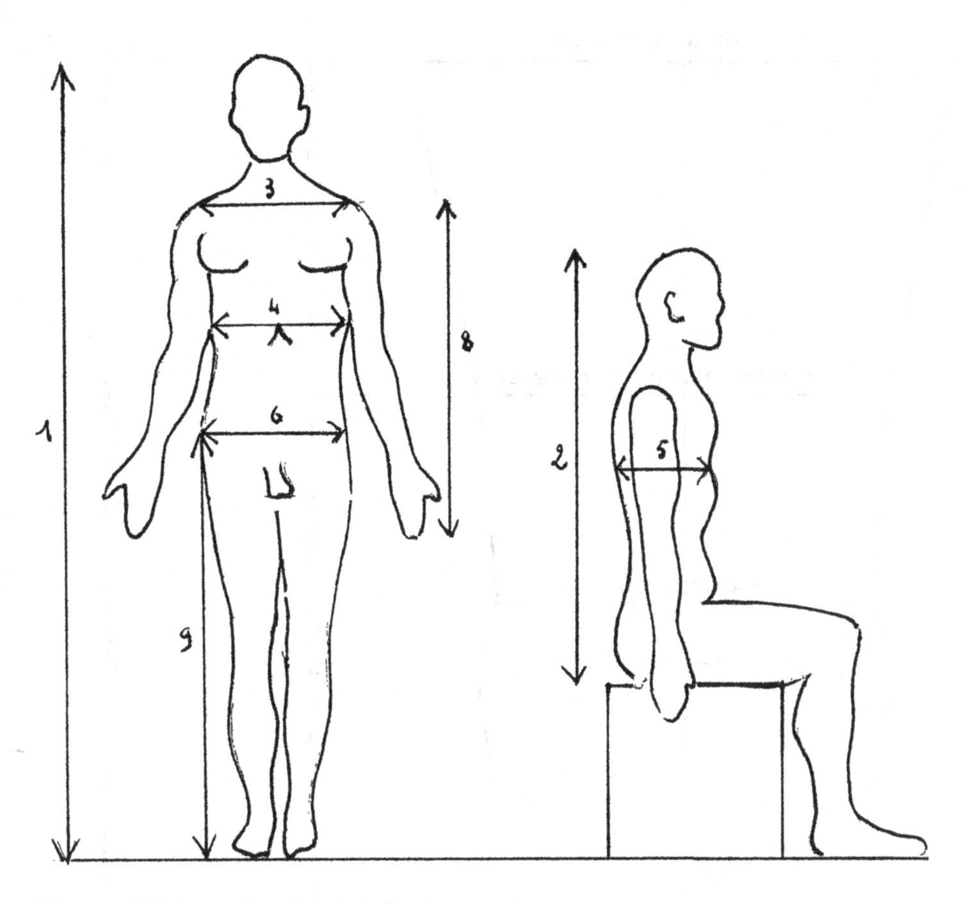

Figure 1. Anatomical characters.

1 Full height 2 Seating height
3 Bi-acromial distance
4 Transversal thoracic diameter
5 Thoracic depth
6 Bi-iliac distance
7 Thoracic member length
8 Pelvic member length

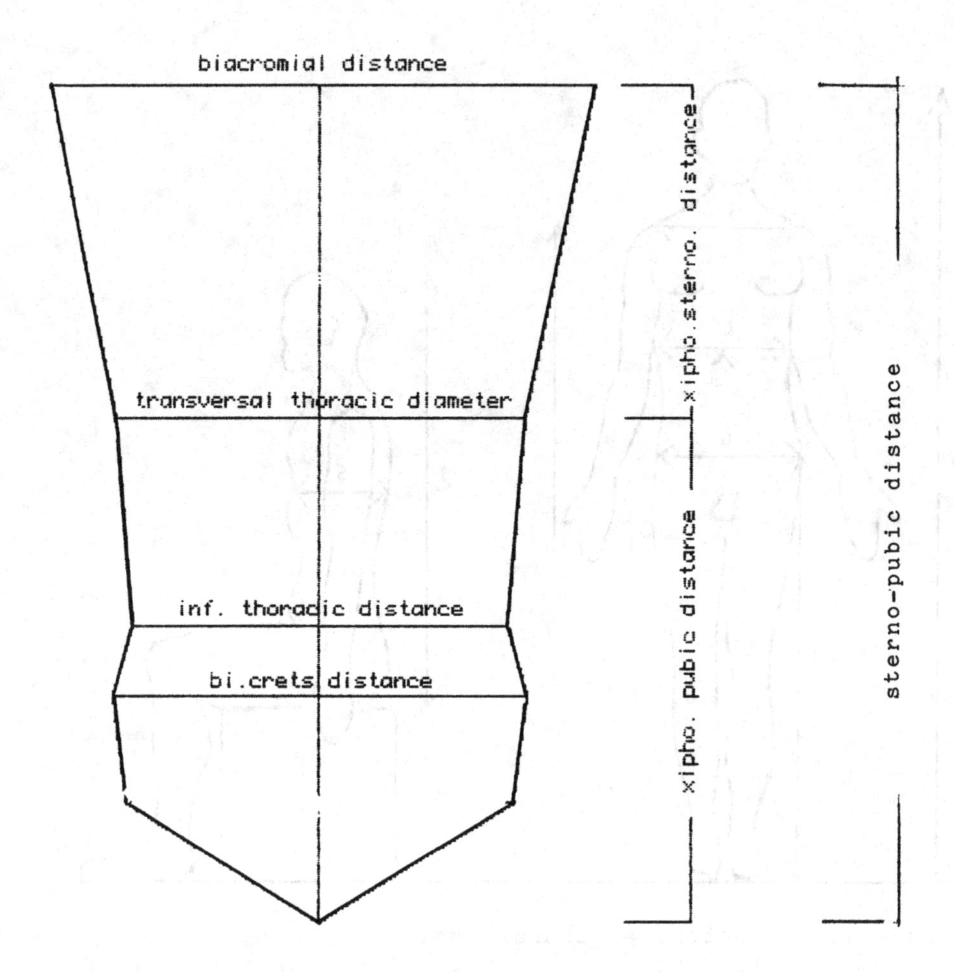

Figure 2. Morphological polygon.

either of the anatomical parameters or of the groups introduce some future prospects in connection with the fuzzy modeling.

2. METHODOLOGY.

2.1. Membership functions of anatomical parameters.

There are many guidelines on developping the membership functions of fuzzy subsets, as surveyed by DUBOIS and PRADE(1980) The sets based on statistics are some of the most naturally fuzzy sets that can be used. The data of each anatomical parameter follow a gaussian distribution as tested by the χ^2 test of fittness. The problem is how to define a membership function from the probability distribution of the data.

In order to construct a reasonable membership function, there are certain conditions which can be imposed to make the fuzzy subset have properties consistent with the user's subjective judgement and the underlying probability distribution From a heuristic viewpoint the data which are most likely have high membership values. However the subset should be as selective as possible. These requirements are quantitatively so described: 1) $E\{\mu(x)|\{x\}$ follow the distribution$\} \geq c$, where the confidence level c should be close to unity. The average membership value assigned to those data distributed according to the density of probability p(x) should be large.

2) $\mu:\{x\}\rightarrow[0,1]$, closed interval of the membership values.

3) $\int\mu^2(x)dx$ should be minimized. This integral is related to the selectivity of the membership function as showed by CZOGALA, GOTTWALD and PEDRYCZ (1982).

The optimal membership function defined by these requirements can be derived using constrained optimization techniques for infinite dimensional spaces (LUENBERGER, 1969). We have to minimize the cost functional f such as:

$$f(\mu) = \frac{1}{2}\int_R \mu^2(x)\,dx \quad \text{with the constraints:}$$

$$G(\mu) = c - E(\mu) = c - \int_R \mu(x)p(x)dx \leqq 0$$

and $$\mu\in\Omega = \{\mu \mid 0 \leqq \mu(x) \leqq 1\}$$

X is the body of the real numbers, Ω is a convex subset of the piecewise continuous functions. f is a convex functional. G maps Ω into P, the closed, non empty cone of the non negative real numbers; G is convex and regular because $G(1)\leqq 0$ by hypothesis. Thus we show that all the conditions hold for the application of the KUHN-TUCKER theorem of the saddle point. See KARMANOV (1977).

The LAGRANGIAN of this problem is:

$L(\lambda,\mu) = f(\mu) + \lambda G(\mu)$. $\lambda \geqq 0$ is the LAGRANGE multiplier.The LAGRANGIAN,linear combination of convex functionals,is itself convex with respect to μ.Thus the necessary , sufficient condition for a function μ* to minimize L is:

$<L'(\mu^*,\lambda),\mu-\mu^*> \geqq 0$, where the quantity $L'(\mu^*,\lambda)$ is the GATEAUX derivative of $L(\lambda,\mu)$ calculated at μ* in the direction of (μ - μ*).The condition is also written as:

$$\int_R (\mu^*(x)-\lambda p(x))(\mu(x)-\mu^*(x))dx \geqq 0$$

The optimal solution μ* is given by GZOGALA and al(1982) and

$$L(\mu^*,\lambda)=\frac{1}{2}\int_R\{I(\lambda p(x))(\lambda p(x)-1)^2-\lambda^2 p^2(x)\}dx + \lambda c$$

with $I(u)=0$ if $u\leq 1$ and 1 else.

CIVANLAR et TRUSSEL (1986) show that the λ* value maximizing the LAGRANGIAN value $L(\mu^*,\lambda)$ is given by the equation:

$$C(\lambda^*)=\int_R\{I(\lambda^* p(x))p(x)+[1-I(\lambda^* p(x))]\lambda^* p^2(x)\}dx-c=0.$$

The same authors show that for every probability density function,there exists a lower bound for the confidence level, over which the optimal membership function μ* satisfies the consistency principle.If the membership value is used as degree of possibility,this consistency principle can be written as:

$$\max_D \mu^*(x)/\max \mu^*(x) \geqq \int_D p(x)dx \quad ,$$ where D is anyone range of the probability distribution.The inequalty means that the degree of possibility of an event is greater than or equal to its degree of probability.

From these results and from the gaussian distribution of $(\bar{x},\sigma)$parameters underlying the data for anyone anatomic parameter,we can use the optimal membership function as:

$\mu^*(z) = \exp[-(z^2-\alpha^2)/2]$ with $\alpha\simeq 0,75$ at the confidence level $c=0,95$ and $z = (x-\bar{x})/\sigma \mid \lambda^* p(z)<1$;

Elsewhere: $\mu^*(z) = 1$ with $z\mid \lambda^* p(z)\geqq 1$;

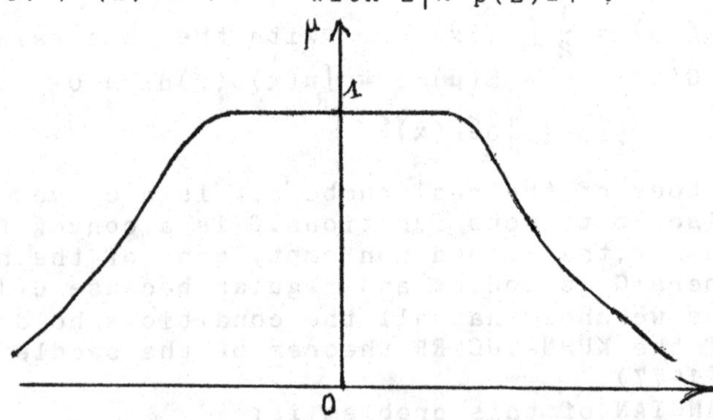

Figure 3. Optimal membership function

2.2. Membership functions of groups.

Each anatomical parameter is viewed as a fuzzy subset A that acts as a fuzzy restriction on the possible values u of the variable X defined on the universe U. A induces a possibility distribution Π_X on the values of X through its membership function μ_A: (ZADEH, 1978)

$$\Pi_X(u)=\mu_A(u)$$

The anatomical parameters we keep are interdependent. The dependence between any two parameters is of unknown nature. According to DUBOIS and PRADE (1980), PRADE (1982), it can be expressed in the following way. If X_1 takes his values on A_1, X_2 then takes his values on A_2 with a degree of possibility so called conditional possibility $\Pi_{(X_2|X_1)}$ and the restriction on the values of X_2 hold through the membership function of the fuzzy implication $A_1 \rightarrow A_2$:

$$\Pi_{(X_2|X_1)}(u_2,u_1) = \mu_{A_1 \rightarrow A_2}(u_1,u_2)$$

According to the works of MIZUMOTO, FUKAMI, TANAK (1979) and ZADEH (1980), HISDAL (1981), PRADE (1982), we can give the general relation between the conditional possibility and the concomitant one $\Pi_{(X_1,X_2)}$ of both X_1 and X_2 taking at one and the same time their values on A_1 and A_2:

$$\Pi_{(X_1,X_2)}(u_1,u_2) = T(\Pi_{(X_2|X_1)}(u_2,u_1),\Pi_{X_1}(u_1)),$$

where T is the triangular norm.

From this relation we have by projection on U_1:

$$\Pi_{X_2}(u_2) = \sup_{U_1} [T(\Pi_{(X_2|X_1)}(u_2,u_1),\Pi_{X_1}(u_1))]$$

Using this equation, we obtain $\Pi_{(X_2|X_1)}$ from both Π_{X_1} and Π_{X_2} :

$$\Pi_{(X_2|X_1)}(u_2,u_1)=\sup_{x\in[0,1]}[T(x,\Pi_{X_1}(u_1))\leq\Pi_{X_2}(u_2)]$$

With the triangular norm $T(a,b)=\max(0,a+b-1)$, $0\leq a,b\leq 1$, we have the following solutions:

$$\Pi_{(X_2|X_1)}(u_2,u_1)=\min(1,1-\Pi_{X_1}(u_1)+\Pi_{X_2}(u_2))$$

$$\Pi_{(X_1,X_2)}(u_1,u_2)=\max(0,\Pi_{(X_2|X_1)}(u_2,u_1)+\Pi_{X_1}(u_1)-1)$$

This choice of the triangular norm gives the largest fuzzy subset A_2 as solution in the sense of inclusion and consequently the more realistic solution $\mu_{A_1 \rightarrow A_2}$ (PRADE, 1980).

We extend the problem to the case of three fuzzy subsets A_1, A_2, A_3 with the conditional possility $\Pi_{(X_3|(X_1,X_2))}$ restric ting the values of X_3 through the membership function of the fuzzy implication: $(A_1,A_2)\to A_3$. This fuzzy implication is expressible as: if X_1 is A_1 and X_2 is A_2 then X_3 is A_3. We have:

$$\Pi_{(X_3|(X_1,X_2))}(u_3,u_1,u_2)=\mu_{(A_1,A_2)\to A_3}(u_1,u_2,u_3)$$

In the same way as before:

$$\Pi_{(X_1,X_2,X_3)}(u_1,u_2,u_3)= \ldots\ldots$$

$$\ldots\ldots T(\Pi_{(X_3|(X_1,X_2))}(u_3,u_1,u_2),\Pi_{(X_1,X_2)}(u_1,u_2))$$

and

$$\Pi_{X_3}(u_3)= \ldots\ldots$$

$$\ldots\ldots \sup_{U_1 x U_2} T(\Pi_{(X_3|(X_1,X_2)}(u_3,u_1,u_2),\Pi_{(X_1,X_2)}(u_1,u_2))$$

From this last equation:

$$\Pi_{(X_3|(X_1,X_2))}(u_3,u_1,u_2)= \ldots\ldots$$

$$\ldots\ldots \sup_{x\in[0,1]} (T(x,\Pi_{(X_1,X_2)}(u_1,u_2))\leq\Pi_{X_3}(u_3))$$

With the same triangular norm:

$$\Pi_{(X_3|(X_1,X_2))}(u_3,u_1,u_2)= \ldots\ldots\ldots$$

$$\ldots\ldots \min(1,1-\Pi_{(X_1,X_2)}(u_1,u_2)+\Pi_{X_3}(u_3))$$

$$\Pi_{(X_1,X_2,X_3)}(u_1,u_2,u_3)= \ldots\ldots$$

$$\ldots\ldots \max(0,\Pi_{(X_3|(X_1,X_2))}(u_3,u_1,u_2)+\Pi_{(X_1,X_2)}(u_1,u_2)-1)$$

We have expressed the algorithm giving step by step the membership value of each fuzzy group (sportsmen, nosportsmen and general population) from all the fuzzy anatomical parameters of this group. Each membership function is normalized

3. RESULTS.

We represent on the table I the mean and the value of the heighest group's membership for each anatomic measurement either of the sportsmen or of the nosportsmen. The drawing's scale is the one from the general population (distribution's mean $\bar{x}$ and standard deviation σ). The statistical comparisons are those between either the means of two different gaussian distributions or two any values of the corresponding gaussian distributions.

With regard to the mean of weight, there is a signifi-

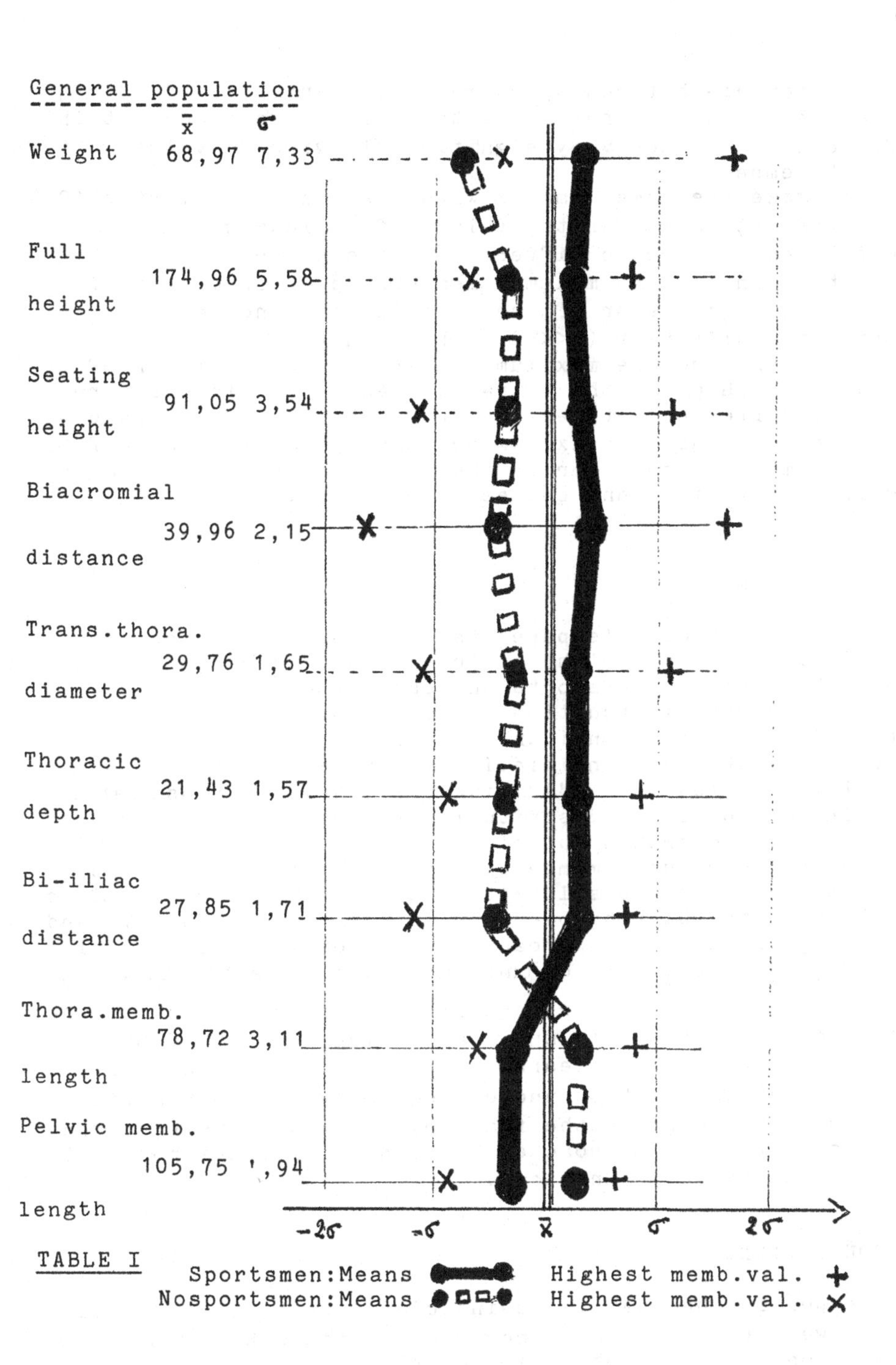
General population
x̄ σ
Weight 68,97 7,33
Full height 174,96 5,58
Seating height 91,05 3,54
Biacromial distance 39,96 2,15
Trans.thora. diameter 29,76 1,65
Thoracic depth 21,43 1,57
Bi-iliac distance 27,85 1,71
Thora.memb. length 78,72 3,11
Pelvic memb. length 105,75 1,94
-2σ
-σ
x̄
σ
2σ
Sportsmen:Means
Highest memb.val.
Nosportsmen:Means
Highest memb.val.

TABLE I

cant difference between sportsmen and nosportsmen ($p<10^{-8}$) . the excess of weight being for the sports group benefit.This difference increases very significantly with the values of the highest membership.

We make the same remarks with the seating height:$p<10^{-5}$ But either the means or the values of maximum membership of the full height are no different from one group to another .

The mean and the maximum membership value either of the biacromial distance or of the bi-iliac distance are very significantly different ($p<<10^{-6}$) between the groups.

The mean and the maximum membership value either of the thoracic depth or of the thoracic transverse diameter are si ficantly different between the groups but only with $p<10^{-2}$.

The mean and the maximum membership value either of the thoracic member lenght or of the pelvic member lenght are not significantly different between the groups.

4. DISCUSSION.

The main object of this paper is the approach of the member ship functions first with the fuzzy anatomic parameters and then with the groups (sportsmen,nosportsmen).

The membership function of each fuzzy anatomical parameter agrees with a confidence level of 95% and so satisfies the consistency principle.It is entirely based on the pro bability distribution underlying the data.Only those data,un derlying a probability distribution,allow such a research of the optimal membership function.

With the group's membership function in view,we consider the conditional possibilities as fuzzy implications.This point is worth thinking over.Any authors as NGUYEN(1978 and 1979) suggest another approach of the conditional possibili ties $\Pi_{(X|Y)}$ or $\Pi_{(Y|X)}$ from the concomitant possibility $\Pi_{(X,Y)}$

We select also the triangular norm corresponding to the fuzzy implication of the multivalent logic of LUCACIEWICZ . This arbitrary choice clears itself only giving the largest fuzzy subset as solution and consequently the most satisfying one with regard to the fuzzy modeling.We have to compare these different approaches of both conditional possibili ties and membership functions.

5. CONCLUSION.

This paper is the starting point of the statistical compari son between the sportsmen and the nosportsmen.We dwell on the foundations of the membership functions.

The fuzzy modeling of the anatomical characters is the first step taking place before the fuzzy modeling of the in dices deduced from these parameters:thoracic index,shoulders breadth index,skelic index,cormic index and others.See SCHR-EIDER (1951,1954,1958),AUBENQUE (1962).

We are convinced that this fuzzy modeling improves the descriptive and comparative abilities of the functional an-thropometry.

6. REFERENCES.

AUBENQUE M. (1962) 'Une statistique des périmètres tho raciques' Biotypologie 23,77-81

BARBARA M. (1929) 'I fondamente della biotipologia uma na' Istuto Editoriale Scientifico,MILANO.

BARROIS X. (1979) 'Etude morphologique comparative d'é tudiants sportifs et non sportifs (à propos de 436 su jets)'.Thèse pour le Doctorat d'Etat en Médecine n 119

BONJEAN P.,ROQUES J.C.,BARROIS X.,LOT J.F.,DUMAS A.,VI GNES J. (1983) 'Vers une Anthropométrie fonctionnel-le' Memo.labo.Anatomie Université Bordeaux II.

CIVANLAR R.,TRUSSEL J. (1984) 'Constructing membership functions using statistical data' Fuzzy Sets and Sys tems 18,1-13.

CZOGALA E.,GOTTWALD S.,PEDRYCZ W. (1982) 'Contribution to applications of Energy measure of fuzzy Sets' Fuz zy Sets and Systems 8,205-214

HISDAL E. (1981) 'A fuzzy "if then else" relation with guaranteed correct inference' in G.E.LASKER ed,Appli ed Systems and Cybernetics,Pergamon Press,vol. VI , 2906-2911

KARMANOV V. (1973) 'Programmation mathématique',1 vol. Edition de Moscou,48-50

MIZUMOTO M.,FUKAMI S.,TANAKA K. (1979) 'Several methods for fuzzy conditional inferences' Proc.18 th IEEE Con ference on Decision and Control,Fort Landerdale,777-782

NGUYEN H.T. (1978) 'On conditional possibility distri-butions ' Fuzzy Sets and Systems 1 299-309

NGUYEN H.T. (1979) 'Towards a calculus of the mathema-tical notion of possibility',in MM.GUPTA,R.K.RAGADE, R.R.YAGER eds.,Advances in Fuzzy Set Theory and App-lications,North-Holland,Amsterdam,235-246

OLIVIER G.,PINEAU H (1961) 'Un morphogramme pour l'é tude des types morphologiques.C.R.Assoc.Anat. NAPLES

PRADE H. (1980) 'Possilité et logique trivalente de LU KASIEWICZ:une remarque' BUSEFAL n°2,Printemps 1980 , Université Paul Sabatier,Toulouse,55-56

PRADE H. (1982) ' Modéles mathématiques de l' Imprécis

et de l'incertain en vue de l'application au Raisonnement naturel.Thése de Doctorat d'Etat es Sciences, n° 1048,Université Paul Sabatier,TOULOUSE.
SCHREIDER E. (1951) 'Analyse factorielle de quelques caractères susceptibles de définir la structure du corps'Biotypologie 12,26-32
SCHREIDER E. (1954) 'Les types humains,méthodes,résultats,concepts',L'Evolution psychiatrique',n° 3,539-556
SCHREIDER E. (1956) 'Morphologie et physiologie',Bull. de l'Inst.Nat.d'Orientation professionnelle 12,1-12
ZADEH L.A. (1978) 'Fuzzy Sets as a basis for a theory of possibility,Fuzzy Sets and Systems 1-1,3-28
ZADEH L.A. (1980) 'Fuzzy Sets versus probability',Proc of the IEEE,Vol 68,421-425

DETERMINING THE EXPECTED VALUE OF A VARIABLE ON THE BASIS OF FUZZY EVIDENCE

Stefan Chanas and Bronisław Florkiewicz
Technical University of Wrocław
Wybrzeże Wyspiańskiego 27
50-372 Wrocław
Poland

ABSTRACT. The problem of determining the expected value of a variable basing on a fuzzy evidence of the type "Pr(V is A) is Q" is considered. Formal properties of the problem, as well as a set of procedures solving it, are given. The procedures are written in TURBO Pascal. With their help r-levels of the fuzzy expected value for any fixed $r \in (0,1]$, as well as for all $r \in (0,1]$ at the same time, can be determined.

1. FORMULATION OF THE PROBLEM

A variable V is given which can assume values in the finite set $X=\{x_1,...,x_n\}$. A fuzzy evidence of the form

$$F=Pr(V \text{ is } A) \text{ is } Q, \qquad (1)$$

where A is a fuzzy subset of X and Q is a fuzzy probability (a fuzzy subset of the unit interval), is the only information about the variable V. The following question arises: what one can say about the expected value of V on the basis of F?

The problem was stated for the first time by Yager [3]. But the solution of the problem presented in [3] seems to be not felicitous.

The weakest point of Yager's approach is lack of a precise definition for a concept of fuzzy expected value of the variable V under the assumption (1). Such a definition should be formulated still before presenting a method for determination of this value.

Another weak point of Yager's work [3] is presentation of his method for a discrete fuzzy probability Q, although he assumes in the introduction (and this assumption should be absolutely maintained) that Q is a fuzzy set in the interval [0,1], and not its certain discrete subset.

Here we present an approach to a solution of the stated problem which is quite different from that proposed by Yager.

W. H. Janko et al. (eds.), Progress in Fuzzy Sets and Systems, 47–62.

2. FORMALIZATION AND SOLUTION OF THE PROBLEM

Let us introduce the notations: $\mu_A(x_i)=a_i$, $i=1,...,n$, and

$$S_n=\left\{p\,|\,p=(p_1,...,p_n),\ p_i\geq 0,\ i=1,...,n,\ \sum_{i=1}^{n} p_i=1\right\}. \tag{2}$$

Each element $p=(p_1,...,p_n)\in S_n$ determines a certain probability distribution over X according to the rule $Pr(V=x_i)=p_i$, $i=1,...,n$.

In [4] Zadeh has suggested that a statement of the form (1) induces in S_n a fuzzy set of possible probability distributions of V under the evidence F. The membership function of this set, which we denote by P/F, is defined as follows:

$$\mu_{P/F}(p)=\mu_Q\left(\sum_{i=1}^{n} a_i p_i\right),\quad p\in S_n. \tag{3}$$

A fuzzy set P/F will be called a fuzzy probability distribution of the variable V under the evidence F.

For the determined universe X the expected value of the variable V is a function of the probability distribution $p=(p_1,...,p_n)$ of this variable:

$$E(V/p)=\sum_{i=1}^{n} x_i p_i,\quad p\in S_n. \tag{4}$$

If we generalize this function according to the extension principle of Zadeh (see e.g. [5]), we will obtain the following natural definition of the fuzzy expected value of V under the evidence F.

Definition 1. A fuzzy set in R, denoted by E(V/F), with the following membership function:

$$\mu_{E(V/F)}(y)=\begin{cases}\sup\limits_{p\in P_y}\mu_{P/F}(p) & \text{if } P_y\neq\emptyset,\\ 0 & \text{otherwise},\ y\in R,\end{cases} \tag{5}$$

where

$$P_y=\left\{p\,|\,p\in S_n,\ \sum_{i=1}^{n} x_i p_i=y\right\},\quad y\in R, \tag{6}$$

is called a fuzzy expected value of the variable V under the evidence F.

Now, the problem consists in determining the r-level of the fuzzy set E(V/F), i.e. the set

$$E(V/F)_r=\{y|\mu_{E(V/F)}(y)\geq r\}, \quad r\in(0,1], \tag{7}$$

for any determined $r\in(0,1]$ as well as for the whole range of variation of r.

If we assume that the fuzzy probability Q is a normal convex fuzzy set in [0,1], with a membership function μ_Q being upper semi-continuous over [0,1], then the r-level of Q is for any $r\in(0,1]$ a closed interval

$$Q_r=\{t|\mu_Q(t)\geq r\}=[q_L(r),q_R(r)], \quad r\in(0,1]. \tag{8}$$

The following theorem, which is of great importance for the approach presented in this work, is valid*).

Theorem 1. For any $r\in(0,1]$ the following statements are true:
(i) $E(V/F)_R\neq\emptyset$ if and only if

$$D_r=\left\{p|p\in S_n,\ q_L(r)\leq\sum_{i=1}^{n}a_ip_i\leq q_R(r)\right\}\neq\emptyset. \tag{9}$$

(ii) If $E(V/F)_R\neq\emptyset$, then

$$E(V/F)_r=[e_L(r),e_R(r)], \tag{10}$$

where

$$e_L(r)=\min_{p\in D_r}\sum_{i=1}^{n}x_ip_i \quad \text{and} \quad e_R(r)=\max_{p\in D_r}\sum_{i=1}^{n}x_ip_i. \tag{11}$$

It results from the Theorem 1 that to determine an r-level of fuzzy expected value $E(V/F)_r=[e_L(r),e_R(r)]$, $r\in(0,1]$, one should solve the following optimization problem:

$$\sum_{i=1}^{n}x_ip_i \rightarrow \min\ (\max), \tag{12}$$

$$x\in D_r:\begin{cases}\sum_{i=1}^{n}p_i=1, & (13)\\ q_L(r)\leq\sum_{i=1}^{n}a_ip_i\leq q_R(r), & (14)\\ p_i\geq 0,\ i=1,\ldots,n. & (15)\end{cases}$$

---

*) Proofs of all the theorems presented in this paper may be found in [1].

Then, $e_L(r)$ is equal to a minimal value of the objective function (12) and $e_R(r)$ to a maximal value of this function. If the problem (12)-(15) has no solution then $E(V/F)_R=\emptyset$. The problem (12)-(15) for the fixed $r\in(0,1]$ is a common linear programming problem, whereas at the whole range of variation of the parameter r the problem (12)-(15) becomes the parametric linear programming problem. Therefore to solve the problem (12)-(15) one might use classical methods of linear programming (see e.g. [2]). However, employing a specific form of the problem one can develop simpler and more efficient solution algorithms. In the following theorems there are presented some properties of the problem (12)-(15) useful for construction of such algorithms.

From this moment we assume (which does not limit the generality of the discussion) that the elements of the set X are increasingly ordered, i.e. the following condition is satisfied:

$$x_1<x_2< \dots <x_n. \tag{16}$$

Theorem 2. For any $r\in(0,1]$ $D_r=\emptyset$ if and only if

$$q_L(r) > \max_{1\le i\le n} a_i \quad \text{or} \quad q_R(r) < \min_{1\le i\le n} a_i. \tag{17}$$

Theorem 3. For any $r\in(0,1]$ the following two statements are true:

(i) If $q_L(r)\le a_1\le q_R(r)$ then $p^*=(p_1^*,\dots,p_n^*)$, where $p_1^*=1$ and $p_i=0$ for $i\neq 1$, is a minimal solution of the problem (12)-(15) (i.e. the optimal solution under the minimization criterion).

(ii) If $q_L(r)\le a_n\le q_R(r)$ then $p^*=(p_1^*,\dots,p_n^*)$, where $p_n^*=1$ and $p_i^*=0$ for $i\neq n$, is a maximal solution of the problem (12)-(15) (i.e. the optimal solution under the maximization criterion).

Theorem 4. For any $r\in(0,1]$ the following statements are true:

(i) If $a_1<q_L(r)$ (or $a_1>q_R(r)$) and $p^*=(p_1^*,\dots,p_n^*)$ is a minimal solution of the problem (12)-(15), then

$$\sum_{i=1}^{n} a_i p_i^*=q_L(r) \quad (\text{or } q_R(r)).$$

(ii) If $a_n<q_L(r)$ (or $a_n>q_R(r)$) and $p^*=(p_1^*,\dots,p_n^*)$ is a maximal solution of the problem (12)-(15) then

$$\sum_{i=1}^{n} a_i p_i^* = q_L(r) \quad (\text{or } q_R(r)).$$

It results from the theorem 4 that if the assumptions of the statements (i)-(ii) are satisfied then the problem (12)-(15) may be replaced by a simpler problem of linear programming:

$$\sum_{i=1}^{n} x_i p_i \to \min\ (\max), \tag{18}$$

$$\sum_{i=1}^{n} p_i = 1 \tag{19}$$

$$\sum_{i=1}^{n} a_i p_i = q(r) \tag{20}$$

$$p_i \geq 0, \quad i=1,\ldots,n, \tag{21}$$

in which $q(r)=q_L(r)$ or $q(r)=q_R(r)$ according to the occuring case.

The next theorem points out important properties of the problem (18)-(21).

Theorem 5. Let $r\in(0,1]$ be any fixed number such that

$$\min_{1\leq i\leq n} a_i \leq q(r) \leq \max_{1\leq i\leq n} a_i \tag{22}$$

and let s and t be indices such that

$$a_s < a_t \quad \text{and} \quad a_s \leq q(r) \leq a_t. \tag{23}$$

(i) If $\hat{p}=(\hat{p}_1,\ldots,\hat{p}_n)$, where $\hat{p}_s=1$ (or $\hat{p}_t=1$) and $p_i=0$ for $i\neq s$ (or $i\neq t$), is a minimal solution of the problem (18)-(21) while $q(r)=a_s$ (or $q(r)=a_t$) and

$$\frac{x_t-x_s}{a_t-a_s} = \min_{i\in\{i\,|a_i>a_s\}} \left\{ \frac{x_i-x_s}{a_i-a_s} \right\}, \quad \left(\text{or} \max_{i\in\{i\,|a_i<a_t\}} \left\{ \frac{x_t-x_i}{a_t-a_i} \right\}\right), \tag{24}$$

then $p^*=(p_1^*,\ldots,p_n^*)$, where

$$p_s^*=\frac{a_t-q(r)}{a_t-a_s}, \quad p_t^*=\frac{q(r)-a_s}{a_t-a_s}, \quad p_i=0 \quad \text{for } i\neq s,t, \tag{25}$$

is a minimal solution of the problem (18)-(21).

(ii) If $\hat{p}=(\hat{p}_1,...,\hat{p}_n)$, where $\hat{p}_s=1$ (or $\hat{p}_t=1$) and $\hat{p}_i=0$ for $i\neq s$ (or $i\neq t$), is a maximal solution of the problem (18)-(21) while $q(r)=a_s$ (or $q(r)=a_t$) and

$$\frac{x_t-x_s}{a_t-a_s} = \max_{i\in\{i\,|a_i>a_s\}}\left\{\frac{x_i-x_s}{a_i-a_s}\right\} \text{ (or } \min_{i\in\{i\,|a_i<a_t\}}\left\{\frac{x_t-x_i}{a_t-a_i}\right\}\text{)}, \qquad (26)$$

then $p^*=(p_1^*,...,p_n^*)$ defined by (25) is a maximal solution of the problem (18)-(21).

The way of calculating r-levels of fuzzy expected value E(V/F) for a fixed r as well as for all r's is apparent from the theorems 1-5. In [1] there are presented in details two solution algorithms. The first one (non-parametric) is intended for the determination of fixed r-levels and the second one (parametric) is intended for the determination of all r-levels of E(V/F). Here we present a full description of a set of procedures written in TURBO Pascal for IBM PC, which realize the both algorithms.

3. COMPUTER PROGRAMS

RLEVEL is a main procedure appropriate for performing the first algorithm. Using this procedure one may obtain the r-level $E(V/F)_r$=[eLr,eRr] for a given value of r. In the case of $E(V/F)_r=\emptyset$ the variables eLr and eRr are not assigned with any value and the occured situation is signaled by setting value true to the boolean variable EMPTY.

Endpar is a main procedure intended for performing the second algorithm. It makes it possible to obtain one of the ends (the left or the right - depending on a value of the variable left) of interval $[e_L(r),e_R(r)]=E(V/F)_r$, $r\in(0,1]$, as a function on r. So, to obtain the full information about E(V/F) procedure endpar must be initiated twice (under different values of the variable left). The function eL(r) (similarly qR(r)) is piecewise linear with respect to the function $q_L(r)$ (if first=true) or $q_R(r)$ (if first=false), where $q_L(r)$ and $q_R(r)$ are the left and right ends, respectively, of the r-levels of Q, e.i. $Q_r=[q_L(r),q_R(r)]$, $r\in(0,1]$. Information about the subintervals and corresponding coefficients are given in arrays e1,e2, endleft and endright.

In both the procedures the following functions, which may not be changed, are used: max, min, sets and sett. The remaining functions - qL and qR (used in the both procedures) as well as qLOplus, qROplus, maxqL and maxqR (used in procedure endpar) - are relevant if Q in the evidence F is a fuzzy number of the trapezoidal form with membership function

$$\mu_Q(y) = \begin{cases} 1 - \dfrac{qleft-y}{alfa} & \text{for} \quad y\in[qleft-alfa,qleft], \\ 1 & \text{for} \quad y\in[qleft,qright], \\ 1 - \dfrac{y-qright}{beta} & \text{for} \quad y\in[qright,qright+beta], \\ 0 & \text{otherwise} \end{cases}$$

In the case of another form of the fuzzy number Q the functions from the last group must be changed. But it is necessary to retain the names of functions, the number of parameters and the meaning of these parameters. All other parameters needed for description of the fuzzy number Q should be included in the functions as nonlocal (global) variables (similarly as parameters qleft, qright, alfa and beta in the present version of the functions).

Main procedures

```
{************************************************************}
{RLEVEL  Determining the r-level of E(V/F) for a fixed r.}
{                                                          }
{procedure RLEVEL(x,a :Vn;n :integer;r :real;             }
{                  var eLr,eRr :real; var EMPTY :boolean);}
{                                                          }
{        x    - array of possible values of variable V;   }
{        a    - array of membership function values of    }
{               fuzzy set A (a[i] - membership degree of }
{               x[i] in fuzzy set A);                     }
{        n    - number of possible values of V;           }
{        r    - a value of considered level;              }
{     eLr,eRr - the left and right end, respectively, of }
{               the interval being the r-level of E(V/F);}
{       EMPTY - a boolean variable which takes on value  }
{               true if the r-level of E(V/F) is empty   }
{               and value false otherwise.               }
{ Description  Procedure determines the r-level of expec-}
{              ted value of variable V under a fuzzy evi-}
{              dence of the type "F=Pr(V is A) is Q"     }
{              according to the methodology proposed in  }
{              in [1].                                    }
{                                                          }
{ Requirement  In program using procedure RLEVEL the      }
{              following constants, types and functions  }
{              must be declared:                          }
{              const                                      }
{                  mvn - maximal value of n;              }
{              type                                       }
{                  Vn   - array[1..mvn] of real;          }
{              function                                   }
{                  max, min, sett, sets, qL, qR.          }
{__________________________________________________________}
```

```
procedure RLEVEL(x,a:Vn;n:integer;r:real;var eLr,eRr:real;
                    var EMPTY :boolean);
var a1,r3,qLr,qRr,help :real;
    t,s                :integer;
    left               :boolean;
begin { RLevel }
  EMPTY:=false;
  left:=false;
  qLr:=qL(r);
  qRr:=qR(r);
  repeat
   left:=not left;
   if (qLr>max(a,n)) or (qRr<min(a,n))
    then begin
           EMPTY:=true;
           exit
        end;
   if left then  a1:=a[1] else a1:=a[n];
   if (qLr<=a1) and (a1<=qRr)
    then begin
           if left then eLr:=x[1] else eRr:=x[n];
           exit
        end;
   if a1 <qLr
    then begin
           if left then t:=1 else t:=n;
           while a[t]<qLr do
             begin
               s:=t;
               t:=sett(a,x,n,s,left)
             end;
           r3:=a[t]-a[s];
           eRr:=(x[s]*(a[t]-qLr)+x[t]*(qLr-a[s]))/r3;
        end
    else begin
          if left then s:=1 else s:=n;
          while a[s]>qRr do
            begin
              t:=s;
              s:=sets(a,x,n,t,left)
            end;
         eRr:=(x[s]*(a[t]-qRr)+x[t]*(qRr-a[s]))/(a[t]-a[s])
        end;
    if left then eLr:=eRr
  until not left
end;  { RLevel }

{************************************************************}
{ endpar   Determining the r-levels of E(V/F) for all r. }
{                                                            }
{ procedure endpar(x,a :Vn;n :integer;left :boolean;        }
{                 var e1,e2,endleft,endright :Vn;           }
{                 var l:integer;var first,EMPTY:boolean);}
{                                                            }
{      x      - array of possible values of variable V;    }
{      a      - array of membership function values of     }
{               fuzzy set A (a[i] - membership degree of }
{               x[i] in fuzzy set A);                       }
{      n      - number of possible values of V;             }
{      left   - =TRUE if the left ends eL(r) of intervals}
{               [eL(r),eR(r)], being the r-levels of       }
```

```
{                  E(V/F), should be calculated and = FALSE }
{                  if the right ends eR(r) should be calcu- }
{                  lated;                                   }
{       e1, e2,                                             }
{ endleft, endright,                                        }
{    l, first - results interpreted as follows:             }
{                  eL(r) (either eR(r))=e1[i]+e2[i]*qL(r)   }
{                  for r belonging to (endleft[i],          }
{                  endright[i]] and  i=1,2,...,l if first=  }
{                  TRUE, or eL(r) (either eR(r))=e1[i]+     }
{                  +e2[i]*qR(r) for r belonging to          }
{                  (endleft[i],endright[i]] and i=1,2,...,l }
{                  if first=FALSE (qL(r) and qR(r) denote   }
{                  the left and right end, respectively, of }
{                  interval being r-level of Q);            }
{        EMPTY - a boolean variable which takes on value    }
{                  TRUE if the r-levels of E(V/F) are       }
{                  empty for all r from interval (0,1] and  }
{                  value FALSE otherwise.                   }
{ Description Using the procedure one can determine all     }
{               r-levels of expected value of variable      }
{               under a fuzzy evidence of the type "F=Pr(V  }
{               is A) is Q" according to the methodology    }
{               proposed in [1].                            }
{                                                           }
{ Requirement  In program using procedure endpar the        }
{               following constants, types and functions    }
{               must be declared:                           }
{               const                                       }
{                  mvn - maximal value of n;                }
{                  eps - small real number (exactness of    }
{                        calculations);                     }
{               type                                        }
{                  Vn  - array[1..mvn] of real;             }
{               function                                    }
{                  max, min, sett, sets, qL, qR, qLOplus,   }
{                  qROplus, maxqL, maxqR.                   }
{___________________________________________________________}
procedure endpar(x,a :Vn;n :integer;left :boolean;
                 var e1,e2,endleft,endright :Vn;
                 var l :integer;var first,EMPTY :boolean);
var a1,r1,r2,r3,rL,rR,maxai,minai,rLeps,qLo,qRo :real;
    t,s                                         :integer;
    stop                                        :boolean;
begin  { endpar }
  EMPTY:=false;
  maxai:=max(a,n); minai:=min(a,n);
  qLo:=qLOplus; qRo:=qROplus;
  if (qLo>maxai) or (qRo<minai)
    then begin
           EMPTY:=true;
           exit
         end;
  l:=0; first:=false; rL:=0.0;
  if left then a1:=a[1] else a1:=a[n];
  if (qLo<=a1) and (a1<=qRo)
   then begin
          r1:=maxqL(a1); r2:=maxqR(a1);
          if r1>r2 then rR:=r2 else rR:=r1;
          if rR-rL>0.0
           then begin
```

```
                  l:=l+1;
                  endleft[l]:=rL; endright[l]:=rR;
                  if left then e1[l]:=x[1] else e1[l]:=x[n];
                  e2[l]:=0.0;
                  rL:=rR
                end
         end;
  rLeps:=rL+eps;
  if rLeps>=1.0 then exit;
  if a1 <qL(rLeps)
    then begin
           first:=true;
           if left then t:=1 else t:=n;
           repeat
             while a[t]<qL(rLeps) do
               begin
                 s:=t;
                 t:=sett(a,x,n,s,left)
               end;
             rR:=maxqL(a[t]);
             if (rR-rL)>0.0
               then begin
                      l:=l+1;
                      endleft[l]:=rL; endright[l]:=rR;
                      r3:=a[t]-a[s];
                      e1[l]:=(x[s]*a[t]-x[t]*a[s])/r3;
                      e2[l]:=(x[t]-x[s])/r3;
                      rL:=rR
                    end;
             rLeps:=rL+eps;
             stop:=(qL(rLeps)>maxai) or (qR(rLeps)<minai)
                                       or (rLeps>=1.0);
           until stop;
           exit
         end;
  if left then s:=1 else s:=n;
  repeat
  while a[s]>qR(rLeps) do
     begin
       t:=s;
       s:=sets(a,x,n,t,left)
     end;
   rR:=maxqR(a[s]);
   if (rR-rL)>0.0
     then begin
            l:=l+1;
            endleft[l]:=rL; endright[l]:=rR;
            e1[l]:=(x[s]*a[t]-x[t]*a[s])/(a[t]-a[s]);
            e2[l]:=(x[t]-x[s])/(a[t]-a[s]);
            rL:=rR
          end;
   rLeps:=rL+eps;
   stop:=(qL(rLeps)>maxai) or (qR(rLeps)<minai)
                           or (rLeps>=1.0);
  until stop;
  exit
end; { endpar }
```

Fixed functions used in the main procedures

```
{*************************************************************}
{ max   Determining maximum array element.                   }
{                                                             }
{ function max(a: Vn;n :integer) :real;                       }
{                                                             }
{            a - array of membership function values of      }
{                fuzzy set A;                                 }
{            n - number of elements in a.                     }
{ Description  A value of function max is equal to maxi-     }
{              mum element among n elements of array a.       }
{                                                             }
{ Requirement  Function is used in procedures RLEVEL and     }
{              endpar.                                        }
{_____________________________________________________________}
function max(a:Vn;n:integer):real;
  var i      :integer;
      s      :real;
  begin { max }
    s:=-1.0e10;
    for i:=1 to n do
      if a[i]>s then s:=a[i];
    max:=s
  end; { max }

{*************************************************************}
{ min   Determining minimum array element.                   }
{                                                             }
{ function min(a: Vn;n :integer) :real;                       }
{                                                             }
{            a - array of membership function values of      }
{                fuzzy set A;                                 }
{            n - number of elements in a.                     }
{ Description  A value of function min is equal to mini-     }
{              mum element among n elements of array a.       }
{                                                             }
{ Requirement  Function is used in procedures RLEVEL and     }
{              endpar.                                        }
{_____________________________________________________________}
function min(a:Vn;n:integer):real;
  var i      :integer;
      s      :real;
  begin { min }
    s:=1.0e10;
    for i:=1 to n do
      if a[i]<s then s:=a[i];
    min:=s
  end; { min }

{*************************************************************}
{ sets Determining index s fulfiling the second part of      }
{      formula (24) (either of formula (26)).                 }
{                                                             }
{ function sets(a,x :Vn; n,t : integer; left : boolean)      }
{              :integer;                                      }
{                                                             }
{     a   - array of membership function values of fuzzy     }
{           set A;                                            }
{     x   - array of possible values of variable V;          }
```

```
{     n   - number of possible values of V;                  }
{     t   - index occuring in formula (24) (either (26));}
{     left- = true if formula (24) is taken into account }
{           and false in the case of formula (26);        }
{ Description  A value of the function sett is equal to  }
{              index s fulfiling the second part of       }
{              formula (24) or (26) (depending on value   }
{              on variable left).    }
{                                                          }
{ Requirement  Function is used in procedures RLEVEL and }
{              endpar.                                     }
{__________________________________________________________}
function sets(a,x:Vn;n,t:integer;left"boolean):integer;
  var i,s      :integer;
    r1,ai,r2   :real;
  begin { sets }
    if left then r1:=-1.0e10 else r1:=1.0e10;
    for i:=1 to n do
      begin
        ai:=a[t]-a[i];
        if ai>eps then
          begin
            r2:=(x[t]-x[i])/ai;
            if left
              then begin
                     if r2>r1 then
                       begin
                        r1:=r2;
                         s:=i
                       end;
                   end
              else begin
                     if r2<r1 then
                     begin
                       r1:=r2;
                       s:=i
                     end;
                   end
          end
      end;
    sets:=s
  end; { sets }

{**********************************************************}
{ sett Determining index s fulfiling the first  part of  }
{      formula (24) (either of formula (26).              }
{                                                          }
{ function sett(a,x :Vn; n,s :integer; left :boolean)     }
{               :integer;                                  }
{                                                          }
{          a   - array of membership function values of }
{                fuzzy set A;                              }
{          x   - array of possible values of variable V;}
{          n   - number of possible values of V;          }
{          s   - index occuring in formula (24) (either }
{                (26));                                    }
{          left- = true if formula (24) is taken into     }
{                account and false in the case of formu-}
{                la (26);                                  }
{ Description  A value of the function sett is equal to  }
{              index t fulfiling the first part of        }
```

```
{               formula (24) or (26) (depending on value  }
{               of variable left).                        }
{                                                         }
{ Requirement   Function is used in procedures RLEVEL and }
{               endpar.                                   }
{_________________________________________________________}
function sett(a,x:Vn;n,s:integer;left:boolean):integer;
  var i,t       :integer;
      r1,ai,r2  :real;
  begin { sett }
    if left then r1:=1.0e10 else r1:=-1.0e10;
    for i:=1 to n do
      begin
        ai:=a[i]-a[s];
        if ai>eps then
          begin
            r2:=(x[i]-x[s])/ai;
            if left
              then begin
                     if r2<r1 then
                       begin
                         r1:=r2;
                         t:=i
                       end;
                   end
              else begin
                     if r2>r1 then
                       begin
                         r1:=r2;
                         t:=i
                       end;
                   end
          end
      end;
    sett:=t
  end; { sett }
```

<u>Functions depending on the assumed form of Q (actual for trapezoidal form of Q)</u>

```
{*********************************************************}
{ qL      The left end of the r-level of Q.               }
{                                                         }
{ function qL(r :real) :real;                             }
{                                                         }
{           r - a real number from the the interval (0,1]. }
{ Description A value of function qL is equal to the left}
{             end of the interval being the r-level       }
{             of the trapezoidal fuzzy number Q=(qleft,   }
{             qright,alfa,beta).                          }
{                                                         }
{ Requirement  Function is used in procedures RLEVEL and }
{              endpar. Variables qleft, qright, alfa,     }
{              beta are not local.                        }
{_________________________________________________________}
function qL(r :real) :real;
  var r1      :real;
  begin { qL }
```

```
    r1:=qleft-alfa*(1.0-r);
    if r1<0.0 then r1:=0.0;
    qL:=r1
  end; { qL }

{**********************************************************}
{ qR      The right end of the r-level of Q.               }
{                                                          }
{ function qR(r :real) :real;                              }
{                                                          }
{           r - a real number from the the interval (0,1]. }
{ Description A value of function qR is equal to the       }
{             right end of the interval being the r-level}
{             of the trapezoidal fuzzy number              }
{             Q=(qleft,qright,alfa,beta).                  }
{                                                          }
{ Requirement  Function is used in procedures RLEVEL and   }
{              endpar. Variables qleft, qright, alfa,      }
{              beta are not local.                         }
{__________________________________________________________}
function qR(r :real) :real;
  var r2     :real;
  begin { qR }
    r2:=qright+beta*(1.0-r);
    if r2>1.0 then r2:=1.0;
    qR:=r2
  end; { qR }

{**********************************************************}
{ maxqL    Maximum r for the left end of r-level.          }
{                                                          }
{ function maxqL(at :real) :real;                          }
{                                                          }
{          at - a real number from the support of fuzzy    }
{               number Q.                                  }
{ Description  A value of function maxqL is equal to       }
{              max r,  s.t. qL(r)<=at, where qL(r) means   }
{              the left end of interval being the          }
{              r-level of trapezoidal fuzzy number         }
{              Q=(qleft,qright,alfa,beta).}                }
{                                                          }
{ Requirement  Function is used in procedure endpar.       }
{              Variables qleft, qright, alfa, beta are     }
{              not local.                                  }
{__________________________________________________________}
function maxqL(at :real) :real;
  var  r1    :real;
  begin { maxqL }
    r1:=(at-qleft)/alfa+1.0;
    if r1>1.0 then r1:=1.0;
    maxqL:=r1
  end; { maxqL }

{**********************************************************}
{ maxqR    Maximum r for the right end of r-level set.     }
{                                                          }
{ function maxqR(as :real) :real;                          }
{                                                          }
{          as - a real number from the support of fuzzy    }
{               number Q.                                  }
{ Description  A value of function maxqR is equal to max   }
```

```
{                r,  s.t. qR(r)>=as, where qR(r) means the }
{                right end of interval being the r-level   }
{                of trapezoidal fuzzy number Q=(qleft,     }
{                qright,alfa,beta).                        }
{                                                          }
{ Requirement    Function is used in procedure endpar.     }
{                Variables qleft, qright, alfa, beta are   }
{                not local.                                }
{__________________________________________________________}
function maxqR(as :real) :real;
  var  r2    :real;
  begin { maxqR }
    r2:=1.0-(as-qright)/beta;
    if r2>1.0 then r2:=1.0;
    maxqR:=r2
  end;  { maxqR }

{**********************************************************}
{ qL0plus     Determining qL(0+).                          }
{                                                          }
{ function qL0plus :real;                                  }
{                                                          }
{ Description    A value of function qL0plus is equal to   }
{                the right hand limit of qL(r) at the point}
{                r=0, where qL(r) means the left end of    }
{                interval being the r-level of trapezoidal }
{                fuzzy number Q=(qleft,qright, alfa,beta)  }
{                (in other words qL0plus is equal to the   }
{                left end of the support of fuzzy numberQ).}
{                                                          }
{ Requirement    Function is used in procedure endpar.     }
{                Variables qleft, qright, alfa, beta are   }
{                not local.                                }
{__________________________________________________________}
function qL0plus :real;
  var  r1    :real;
  begin { qL0plus }
    r1:=qleft-alfa;
    if r1<0.0 then r1:=0.0;
    qL0plus:=r1
  end; { qL0plus }

{**********************************************************}
{ qR0plus     Determining qR(0+).                          }
{                                                          }
{ function qR0plus :real;                                  }
{                                                          }
{ Description    A value of function qR0plus is equal to   }
{                the right hand limit of qR(r) at the point}
{                r=0, where qR(r) means the right end of   }
{                interval being the r-level of trapezoidal }
{                fuzzy number Q=(qleft,qright,alfa,beta)   }
{                (in other words qR0plus is equal to the   }
{                right end of the support of fuzzy number  }
{                Q).                                       }
{                                                          }
{ Requirement    Function is used in procedure endpar.     }
{                Variables qleft, qright, alfa, beta are   }
{                not local.                                }
{__________________________________________________________}
function qR0plus :real;
```

```
var  r2     :real;
begin { qROplus }
  r2:=qright+beta;
  if r2>1.0 then r2:=1.0;
  qROplus:=r2
end; { qROplus }
```

4. NUMERICAL EXAMPLE

Let us assume that in evidence "F=Pr(V is A) is Q" $X=\{1,2,3,4\}$, $A=0/1+0.2/2+1.0/3+0.5/4$ and Q is a triangular fuzzy number in [0,1] with a membership function:

$$\mu_Q(y)=\begin{cases} 0 & \text{for } 0\le y\le 0.4, \\ 1-(0.6-y)/0.2 & \text{for } 0.4\le y\le 0.6, \\ 1-(y-0.6)/0.4 & \text{for } 0.6\le y\le 1.0. \end{cases}$$

Using the RLEVEL procedure we obtain for a few fixed values of r:

$E(V/F)_{0.2}=[1.88,\ 4.0]$,

$E(V/F)_{0.6}=[2.04,\ 3.96]$,

$E(V/F)_{1}=[2.2,\ 3.8]$.

If we apply the endpar procedure, then we obtain for $r\in(0,1]$

$$E(V/F)_r=[e_L(r),\ e_R(r)],$$

where

$$e_L(r)=1+2q_L(r)=1.8+0.4r,\ r\in(0,1]$$

and

$$e_R(r)=\begin{cases} 4+0q_L(r)=4 & \text{for } r\in(0,0.5], \\ 5-2q_L(r)=4.2-0.4r & \text{for } r\in(0.5,1]. \end{cases}$$

REFERENCES

[1] Chanas, S. and Florkiewicz, B. (1988) 'Deriving expected values from probabilities of fuzzy subsets', Inst. Org. i Zarz. Politechniki Wr., Report No. 34 (submitted to European Journal of Operational Research).

[2] Gass, S.J.(1969) Linear Programming - Methods and Applications, McGraw-Hill Book Company, New York.

[3] Yager, R.R. (1986) 'Expected values from probabilities of fuzzy subsets', European Journal of Operational Research 25, 336-344.

[4] Zadeh, L.A. (1979) 'Fuzzy sets and information granuality', in M.M. Gupta, R.K., Ragade and R.R. Yager (eds),.Advance in Fuzzy Set Theory and Application, North-Holland Publ. Company, pp. 3-18.

[5] Zadeh, L.A. (1973) The Concept of a Linguistic Variable and its Application to Approximate Reasoning, American Elsvier Publ. Company, New York.

A FUZZY TIMES SERIES ANALYZER

A. GEYER
Department of Operations Research
A. GEYER-SCHULZ and A. TAUDES
Department of Applied Computer Science
Vienna University of Economics and Business Administration
Augasse 2-6
A-1090 Vienna
Austria

ABSTRACT. This paper presents elements of an integrated time series analysis package that relate to the application of fuzzy set theory. We describe some aspects of a fuzzy time series analyzer, its architecture and show how fuzzy time series features can be extracted from the data.

1. Introduction

Time series analysis (TSA) and forecasting may be considered as an important supporting element of planning processes. As a matter of fact there has been made considerable progress in the improvement and development of new analysis techniques over the past ten years (see Newbold [12], [13] and [14] for overviews). At the same time there has been a rapid increase in the availability of both data as well as TSA and forecasting software. Given this environment, it could be assumed that TSA and forecasting have become well established planning tools. However, a closer look at the problems associated with practical TSA and a deeper understanding of the available techniques indicate that this progress does not necessarily translate into an advantage for the practitioner. The following reasons for this deficiency can be mentioned:

1. Unless very simple methods are used the various analysis techniques are difficult to understand, or, it is difficult to understand the implications and limitations incurred by applying a particular technique.
2. Even for experts it is time consuming to find appropriate models even if simple models are applied. This becomes particulary disadvantegeous when a large number of time series has to be dealt with.
3. TSA is inherently fuzzy. There is not a single step in the time series model building process where decisions can be based on unambiguous information. It is this aspect of the problems associated with TSA that will be the major subject of this paper.

The first two problems have led to the development of automatic or semi-automatic TSA systems and TSA expert systems. To overcome most of the typical problems in TSA

W. H. Janko et al. (eds.), Progress in Fuzzy Sets and Systems, 63–74.

the fuzzy time series analyzer to be described in this paper has been developed. It should be emphasized that it is not the purpose of this attempt to replace existing theoretical and practical knowledge by fuzzy sets but rather to provide appropriate support for time series model building. The time series analyzer is a very general tool for TSA which tightly couples TSA software and statistical knowledge bases. The system is open for additions by the user who can add his own methods and heuristics. A fuzzy-rule language is provided to formulate user defined elements of the system.

The paper is organized as follows. Section 2 provides evidence for the fuzziness involved in TSA on the basis of three different approaches. Section 3 reviews automatic methods and expert systems proposed for TSA. The fuzzy time series analyzer is introduced in section 4 and special emphasis is given to the way fuzzy time series features are extracted from the data.

2. Time Series Analysis - A Fuzzy Problem

We will exemplify the fuzziness involved in time series analysis by considering ARMA models as generated by the Box-Jenkins [2] methodology.

The Box-Jenkins methodology is a structured approach to time series model building. The resulting ARMA models are derived in a decision process whereby statistical evidence has to be interpreted and conclusions have to be drawn in each step. The first stage is to identify a proper model. In this context Granger and Newbold ([7] p.115) note that "... the stage that causes the most difficulties in practical attempts at time series model building is identification. Here one is required to choose from a wide class of models a single process that might adequately describe a given time series. While some objective criteria are available on which a rational choice can be based, it remains the case that there does not exist a clearly defined procedure leading in any given situation to a unique identification. Rather, it is necessary to exercise a good deal of judgement at this stage."

In order to support the decisions to be made a number of theoretical relationships can be employed. Typically however, the corresponding rules are inexact verbal statements like " ... the autocorrelation function (ACF) of an autoregressive process of order p *tails off*, its partial autocorrelation function (PAC) has a *cutoff* after lag p" ([2] p.175). The terms *tail off* and *cutoff* describe theoretical relationships derived from time series theory. In the course of model building theoretical knowledge of this kind has to be integrated into the decision process. However, this is further complicated by the fact that statistics derived from the data need not resemble the theoretical patterns to be searched for.

Figure 1 contains theoretical shapes of the two mentioned functions and figure 2 contains estimated shapes resulting from actual time series data. Typical conclusions drawn from such printouts read as follows: "... from the autocorrelation function of z it is seen that after lag 1 the correlations do decrease *fairly* regularly. [...] The partial autocorrelation function *tends* to support this possibility." [2] p.179. Thus, conclusions are almost always uncertain and experienced time series analysts try to confirm their conclusions by checking and cross-checking using other indicators. This of course requires even more experience and entails further decision problems.

Figure 1: Theoretical shapes of ACF and PAC

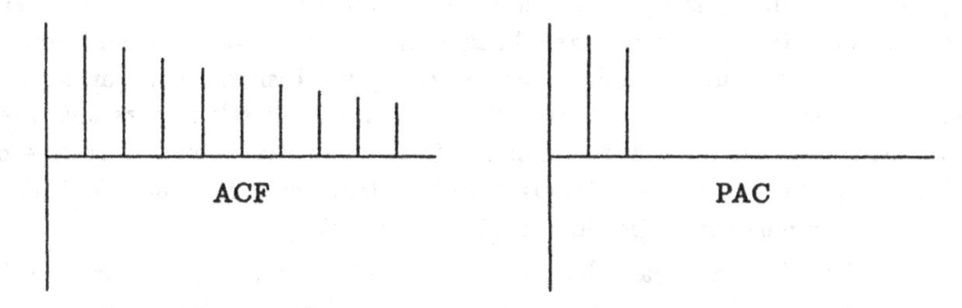

Figure 2: ACF and PAC estimated from actual data

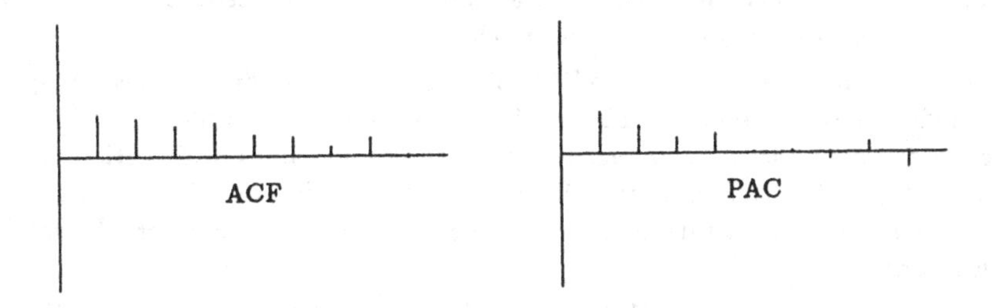

This example shows that it is very difficult to select and build proper time series models. Experience in the field can only be obtained in a long and time consuming process. At the same time it is evident that the knowledge of an expert cannot be easily represented in an expert system. The problem is that automatic methods of model identification and selection require crisp decisions while TSA is inherently fuzzy! Since experience consists of various fuzzy rules the fuzzy time series analyzer was designed to include fuzzy logic in an expert system.

Before presenting the fuzzy time series analyzer in more detail we will briefly review existing approaches to time series modeling by expert systems or automatic methods.

3. Automatic Approaches To Time Series Analysis

It has been often mentioned as a major drawback of the Box-Jenkins methodology that it is not a fully automatic technique. Consequently, a number of approaches have been suggested to overcome this drawback.

CAPRI is a system developed by Libert [11] that builds an appropriate ARMA model from the data. The procedure selects an optimal parameter for the Box-Cox transformation and determines the necessary degree of differencing. In order to detect seasonality of the data "CAPRI proceeds like the modeller interpreting an acf" (p.97) by searching for significant autocorrelations at seasonal lags. The choice from ARMA(p,q) models is confined such that $p + q \leq 3$. Upon comparison of ACF and PAC coefficients at lags $k = 1, \ldots, p + q + 2$ with a *threshold value* possible models are selected. The treshold value is not specified in Libert's papers. Each model in the list of possible models is estimated and the significance of parameters is checked. Insignificant parameters are dropped and the model is reestimated. It is not clear from Libert's papers how the final or "best" model is selected. Further it appears impossible that a multiplicative model can be detected by this procedure. Experimental evidence suggests that CAPRI works as well as a human expert using the Box-Jenkins methodology.

Conceptually the SIFT approach by Hill and Woodworth [9] is similar to CAPRI. Only model orders up to $p = q = 4$ are considered and the model selection is based on inspection of typical correlation properties of the series. Details of the algorithm to determine patterns like cut-off, exponential decay and sine waves are not described in their paper although this is the most difficult and fuzziest step in the model building procedure. Possible models are selected in a "model release procedure" where Akaike's FPE criterion is used to choose the "best" model.

In summary, a basic element of CAPRI and SIFT is to confine the search space of possible models a priori and the model selection is based on similar techniques that involve crisp decisions based on comparisons with threshold values or detection of typical correlation patterns like tail-off or cut-off. The procedures do not take into account the fuzziness of the problem but rather transform the problem such that it can be solved by binary decisions.

The approach by Lee and Park [10] is particulary devoted to pattern recognition. Appropriate ARMA model orders are determined on the basis of the extended sample autocorrelation function suggested by Tsay and Tiao [16]. This function is estimated from the data and compared to a set of predefined theoretical patterns contained in a database. A "decision value" resulting from this classification can be interpreted as a measure of the certainty with which the system has chosen the particular model structure. Lee and Park's approach is not comprehensive since it only considers the class of stationary non-seasonal ARMA models.

A conceptually different approach is Weitz' [17] knowledge based forecasting advisor NOSTRADAMUS. The purpose of this "prototype expert system" is to "assist the non-expert forecasters in the selection and use of appropriate forecasting methods". NOSTRADAMUS considers a variety of well-known forecasting techniques. In an interactive user dialog it attempts to assess whether a particular method is feasible and can be suggested for the series under study. However, since the system is programmed in standard LISP the actual analysis *cannot* be performed by the system. That is, a potential user would have to sit in front of two terminals, carry out the calculations on one and get advice from the other.

The system developed by Eckhardt [4] is also not confined to a single TSA approach but is able to perform all required calculations. It contains a model base of several forecasting methods (especially exponential smoothing models) where each method is assigned a value between 0 and 1 indicating the appropriateness of the model given different forecasting horizons, number of observations, trend patterns and the existence of seasonality. The values assigned are based on "comprehensive comparisons from the literature"(p.122). The properties of a particular time series are identified in a user dialog and from statistical tests. Matching these properties with the suitability indicators allows to present several models to the user, where each model is assigned a value reflecting the degree of appropriateness. Thus, the system supports the user in the decision which model to choose.

Apart from the approaches discussed in this section there exist other automatic or black-box modeling approaches such as Flexicast [3] and TSA software packages such as AUTOCaST [5], AUTOBOX [1] or 4CaST [8] that include automatic modes.

4. The Fuzzy Time Series Analyzer

The fuzzy time series analyzer (FTSA) is the intelligent component of a TSA software package which supports either fully automatic model-building or provides a tutoring component which aids the user in the model building process by giving explanations and recommendations or reviews of user actions. In addition to the FTSA, a TSA tool-kit, a time series model management system and extensive documentation and help facilities are included in the package.

The FTSA integrates a fuzzy rule-based expert system and the TSA software tool-kit by extending the fuzzy rule-language (which is described in detail in Geyer-Schulz [6]) with tests of time series properties and with general transformations of time series including estimation routines. The rule-base consists of rule that correspond to the feature testing, model selection and diagnostic checking elements of the analysis process. The rule-interpreter performs a systematic model building process and thus mimicks the identify-estimate-check cycles as performed by a human time series analyst.

We will proceed in two steps: section 4.1. gives a rough sketch of the rule-based expert system of the FTSA, section 4.2. explains the integration of the TSA components into the fuzzy rule-language using the problem of identifying the correct order of an ARMA(p,q) model as an example.

4.1. THE EXPERT SYSTEM

The expert system has two building blocks: (a) a database which contains rule bases and time series and (b) a virtual machine which comprises the fuzzy rule interpreter and the fuzzy rule compiler.

4.1.1. Database. The database represents the memory structure of the system. The analysis strategy of an expert in TSA is specified in the rule-bases, the observations of the time series, statistics and features like trend or seasonality are stored in the time series objects.

In general rules have the following basic structure:

<action part> IF <condition part>;

In the context of the Box-Jenkins methodology a simple example of a rule is (a complete description of the syntax of rule-bases can be found in [6]):

TRANSFORM_ARMA IF TEST_ARMA IS HIGH;

The action TRANFORM_ARMA estimates an ARMA(p,q) model and returns the residuals if the test procedure TEST_ARMA indicates a high degree of confidence that at least for one ARMA(p,q)-model the ACF and PAC patterns are typical. A possible variant of the procedure TEST_ARMA is described in the next section.

The time series object has the structure of a named association table which allows for symbolic references to the entries. A set of access functions as defined in [6] p.8 are provided which handle all references to the time series and its features.

4.1.2. Virtual Machine. The rule compiler translates a rule-base into its internal representation and resolves name bindings. It binds the time series tests and transformations to the appropriate elements of the runtime evaluation stack of the rule interpreter. For the example rule given above the action part is rewritten as (for a description of rewrite rules see [6] p.59):

(NEW_OBJECT ← TRANSFORM_ARMA OLD_OBJECT)

and the condition part is rewritten as:

(OLD_OBJECT ← TEST_ARMA OLD_OBJECT) IS HIGH

The fuzzy rule interpreter first executes unconditional actions and then invokes a backtrack algorithm on the remaining rules of the rule base. The backtrack algorithm simulates the behaviour of a time series analyst when building a model. It is structurally equivalent to the algorithm in Nilsson [15] p.59, only the predicate functions of the algorithm get their meaning from the context of TSA. *Successful termination* means that a proper model has been established and the algorithm halts. *Dead end* means that the algorithm will not be able to find a proper model on the current path. *Testing a depth bound* ensures that the algorithm will stop. *Return from a failed path* means that another model has to be tried. *No more rules* means that the algorithm has tried all models and has to return to an earlier choice point.

The procedure *apply rules* executes all conditional parts of the rule-base. The result is a time series model possibility function, which is stacked on the rule-stack. The procedure *choose rules* selects the model with the highest possibility value. The action which generates the chosen model is fired by the procedure *execute rules.*

The flow of the algorithm can be set by the statements EXPLAIN, EXPLAIN_ALL and FAIL, respectively. EXPLAIN identifies a proper model, saves it and terminates backtracking. EXPLAIN_ALL enumerates all proper models. FAIL discards fruitless search paths. EXPLAIN_ALL and FAIL set the fail-predicate of the backtrack algorithm.

4.2. INTEGRATION OF TSA COMPONENTS INTO THE FUZZY RULE-LANGUAGE

When integrating TSA components into the fuzzy rule-language two classes of TSA components have to be distinguished: (a) transformations of the time series and (b) test routines. Transformation routines adhere to the following outline: they fetch parameters from the time series object for the numerical transformation routine, call the numerical transformation routine from the TSA tool-kit and build a new time series object with the transformed time series.

The integration of test routines is more complicated since each test routine is modelled according to a feature extraction process. This process returns the required information to the test procedure in three steps: first, calculate a statistic using estimation and/or test functions of the TSA tool-kit. Second, a pattern recognition function uses the statistic to provide information about the existence of a pattern. Third, this information is converted by a translation routine into linguistic expressions. The linguistic expressions must have the same internal representation as the expressions used in the fuzzy rule-language. This structure allows us to experiment with different, even adaptive pattern recognition approaches.

The details of the feature extraction process will now be explained by describing one possible way to identify the order of an ARMA(p,q) model. It should be noted that the described procedure is not considered to be the only or best procedure available but it is chosen (a) to show how the problem described in section 2. may be tackled by using fuzzy set theory and (b) to make the procedure more explicit than it is done in the papers on CAPRI and SIFT. Due to lack of space the reader is assumed to be familiar with details of ARMA model identification.

4.2.1. Statistical Functions. In the Box-Jenkins methodology the statistics used to identify the ARMA model order are ACF and PAC.

4.2.2. Pattern Recognition. The purpose of pattern recognition in the present context is to compare the theoretical patterns of ACF and PAC classified by Box-Jenkins to the patterns estimated from the time series under consideration. This procedure requires (a) the generation of theoretical reference patterns (which would usually be stored in a database), (b) the matching of theoretical and estimated patterns and (c) the aggregation of information using theoretical considerations and heuristics.

4.2.3. Reference Patterns. The patterns we are looking for are verbally described by Box-Jenkins as "damped exponentials", "damped sine waves" or mixtures thereof and cut-offs. There are several possibilities for a functional representation of these patterns. Here we choose functions that are based on theoretical relations for particular ARMA processes.

The ACF of an autoregressive process of order 1 – denoted AR(1) – tails off according to

$$r_k^{(1)} = a_1^k$$

where a_1 is the autoregressive coefficient and $k = 0, 1, \ldots, M$ where M is usually chosen to be 24, 36 or 48. The ACF of an AR(2) process can follow either a mixture of damped exponentials (if the roots of the AR polynomial are real) or damped sine waves as defined by the difference equation

$$r_k^{(2)} = a_1 r_{k-1}^{(2)} + a_2 r_{k-2}^{(2)}$$

where a_1 and a_2 are autoregressive coefficients, $r_0^{(2)} = 1$ and $r_1^{(2)} = a_1/(1 - a_2)$.

Cut-off patterns are represented by

$$r_k^{(3)} = \begin{cases} 1 & k = 1, \ldots, K \\ 0 & k > K \end{cases}$$

K varies from 1 to 5, say, or any other number that may appear reasonable.

Varying the parameters a_1 and a_2 over a practically relevant range allows the construction of some 100 different typical patterns. Although this reference set corresponds only to AR(1) and AR(2) processes it contains most of the important shapes that are also valid for higher order models.

4.2.4. Pattern Matching. The patterns $r_k^{(1)}$, $r_k^{(2)}$ and $r_k^{(3)}$ are regressed against the estimated ACF and PAC using the following linear models:

$$\rho_k w_k = b_0 + b_1 r_k^{(i)} w_k \qquad i = 1, 2$$

$$\phi_k w_k = c_0 + c_1 r_k^{(i)} w_k \qquad i = 1, 2$$

where ρ_k and ϕ_k denote estimated ACF and PAC, b_0, b_1, c_0, and c_1 are regression coefficients and $w_k = (M - k + 1)^2$ and thus emphasizes the importance of ACF and PAC near k=1. For cut-off patterns ($i = 3$) the same regression is used but ρ_k and ϕ_k are replaced by an indicator value which is equal to 1 if $\rho_k > 2/\sqrt{T}$ and 0 otherwise. T is the number of observations and $2/\sqrt{T}$ is the level of significance. Thus, only significant lags are considered as contributing to cut-off patterns. The coefficient of determination R^2 of each regression indicates the degree of correspondence between the actual and the theoretical shape.

4.2.5. Aggregation Of Information. In order to be able to draw inferences the available R^2 values are separated into two groups. One group derived from the ACF regressions and one from the PAC regressions. Each group is again subdivided into those R^2 values corresponding to tail-off patterns ($i = 1, 2$) and cut-off patterns ($i = 3$). The maximum of all R^2 values of the first subgroup indicates the degree of confidence in a tail-off pattern. The maximum R^2 from ACF regressions is denoted by D^A, the one for PAC by D^P. R^2 values of the second subgroup indicate the degree of confidence in a cut-off at lag 1, 2 and so on. These R^2 values are denoted by C_k^A for ACF and C_k^P for PAC (see table 1 for a compilation of the symbols used).

Testing for a pure AR process is performed by calculating

$$P_k = (D^A + C_k^P)/2$$

Table 1: Symbol table

Pattern	Maximum believe values obtained for ACF	PAC
Tail-off	D^A	D^P
Cut-off at lag k	C_k^A	C_k^P
Tail-off after k lags	B_k^A	B_k^P

for each k for which a cut-off has been considered. P_k combines the degrees of confidence in the existence of a tail-off and cut-off pattern and can thus serve as a measure of belief in an AR(k) model. Testing for a pure MA process is performed by calculating

$$Q_k = (D^P + C_k^A)/2$$

for each k for which a cut-off has been considered. Q_k indicates the degree of confidence in an MA(k) model. This simple way of combining evidence does not exploit all available information. The particular pattern $r_k^{(1)}$ or $r_k^{(2)}$ which yields the maximum R^2 could also be used to indicate whether an AR(1) or AR(2) model is appropriate.

The detection of mixed ARMA models is more complicated because the theory admits only few conclusions that can be drawn from estimated patterns in ACF and PAC. In general the ACF of an ARMA(p,q) model tails off after lag $q-p$ lags and the PAC after $p-q$ lags (see [2] p.79). However, this does not allow for a unique identification. For instance a tail-off of both ACF and PAC after lag 1 could correspond to an ARMA(2,1), ARMA(3,2), ...or ARMA(1,2) model and so on. Further an immediate tail-off of both ACF and PAC is consistent with several ARMA(p,p) but the value of p cannot be inferred from the data. If the lag after which a tail-off occurs in ACF is different from the lag in PAC the appropriate model is probably a pure AR or MA rather than an ARMA model.

In the present context we do not require a unique identification. It is sufficient if the available evidence is appropriately transformed into a fuzzy set. This is done by regressing estimated ACF and PAC against a suitable modification of the tail-off function $r_k^{(1)}$ such that the tail-off starts after 1,2,...lags. Denote by B_k^A the maximum R^2 value of the regression on ACF where k indicates the lag after which the tail-off pattern starts. The corresponding maximum for PAC is B_k^P. The degree of confidence in an ARMA(1,1) model is denoted by PQ_{11} and calculated as follows

$$PQ_{11} = (D^A + D^P)/2$$

The PQ values for ARMA(i,j) models is derived from

$$PQ_{i,j} = [(B_k^A + B_k^P)/2 + P_i]/2 \quad \forall i = j+k;\ j = 1,2,\ldots;\ k = 1,2,\ldots$$

and

$$PQ_{i,j} = [(B_k^A + B_k^P)/2 + Q_j]/2 \quad \forall j = i+k;\ i = 1,2,\ldots;\ k = 1,2,\ldots$$

Table 2: Belief values derived from ARMA pattern matching

		autoregressive order				
		0	1	2	3	4
	0	.	0.769	0.995	0.863	0.807
moving	1	0.665	0.983	0.975	0.889	0.857
average	2	0.801	0.878	.	0.909	0.861
order	3	0.906	0.911	0.931	.	0.881
	4	0.989	0.948	0.952	0.972	.

The inclusion of the pure AR and MA model plausibility indicators P_i and Q_j is not based upon theoretical considerations but on a heuristic that attempts to introduce information on the possible predominance of AR or MA terms in the ARMA model. A similar heuristic is not available to indicate the order p of ARMA(p,p) models where $p > 1$. But the principle of parsimony favours low order models and thus ARMA(p,p) models where $p > 1$ are not further considered in this procedure. Table 2 contains the results of the application of the procedure to the first differences of the sales data [2] p.537. The table indicates that a pure AR(2) model is the most plausible, followed by a MA(4) and an ARMA(1,1) model. Given the estimated ACF and PAC for the series (see figure 2) this can be considered as a reasonable model identification.

In order to detect seasonal time series models the same procedure is applied to every second, every third, until every twelvth autocorrelation and partial coefficient depending on the evidence provided by other indicators of seasonality such as the spectrum. Thus, also multiplicative ARMA models can be identified.

4.2.6. Translation Function. Since the expert system expects a linguistic value as a result of the feature extraction process it is necessary to convert the matrix of belief values to the internal representation of a linguistic expression.

The internal representation of a linguistic expression is achieved in the fuzzy rule-based expert system by a tagged data-representation with implicit conversion. A full description of all supported data types and the implicit conversion rules is given in [6] p.30-43. For the present example only the simplest tagged data type is needed. It is a data type for data given on an abstract and normalized 0–1 scale. The APL2 boxed array representation of this data type is e.g.:

```
| S01 | 0.93 |
```

For instance, an ARMA(2,0) model with a belief value of 0.995 is represented with the following tagged data structure:

```
| OF | ARMA 2 0 | | S01 | 0.995 | | NIL | NIL |
```

The translation function converts table 2 to the model possibility function. Being a fuzzy subset on a discrete base set it is represented as a vector of data elements. This model possibility function is the explicit result of the procedure TEST_ARMA which is used by the backtrack algorithm (see section 4.1.).

5. Summary

In this paper we have presented a fuzzy time series analyzer that is the intelligent component of a time series analysis software package. It integrates a fuzzy rule-based expert system and a time series analysis tool-kit. For the problem of identifying the order of ARMA(p, q) models we have demonstrated the details of a pattern recognition method and its integration in the expert system.

References

[1] Automatic Forecasting Systems Inc.(1986), "AUTOBOX User's Guide", Hatboro Pa.

[2] Box G.E.P. and G.M.Jenkins(1976) "Time Series Analysis Forecasting and Control, revised edition", *Holden-Day:San Francisco.*

[3] Coopersmith L.W.(1979) "Automatic forecasting using the FLEXICAST system", *TIMS Studies in the Management Sciences*, **12**, 265-278.

[4] Eckhardt Th.(1980) "Eine Prognose-Methodenbank für Kleinrechner",115-133, J.Schwarze, *Angewandte Prognoseverfahren*, Neue Wirtschaftsbriefe, Berlin.

[5] Gardner Jr. E.S.(1986) "AUTOCaST", Levenbach Associates. Suite 348, 103 Washington St, Morristown, NJ 07960.

[6] Geyer-Schulz A.(1988) "Fuzzy Rule-Based Expert Systems", *APL Techniques in Expert Systems*, eds.J.R.Kraemer, P.C.Berry, ACM SIGAPL, Syracuse NY.

[7] Granger C.W.J. and Newbold P.(1986) "Forecasting Economic Time Series, 2nd edition", *Academic Press:Orlando*, **1**, 67-82.

[8] Heurix Computer Research "4CaST/2", Levenbach Associates. Suite 348, 103 Washington St, Morristown, NJ 07960.

[9] Hill G.W. and D.Woodworth(1980) "Automatic Box-Jenkins forecasting",*J.Opl Res. Soc.*, **31**, 413-422.

[10] Lee K.C. and S.J.Park(1988) "Decision support in time series modeling by pattern recognition", *Decision Support Systems*, **4**, 199-207.

[11] Libert G.(1984) "An automatic procedure for Box-Jenkins model building", *EJOR*, **17**, 95-103.

[12] Newbold P. (1981) "Some recent developments in time series analysis", *International Statistical Review*, **49**, 53-66.

[13] Newbold P. (1984) "Some recent developments in time series analysis - II", *International Statistical Review*, 52, 183-192.

[14] Newbold P. (1988) "Some recent developments in time series analysis - III", *International Statistical Review*, 56, 17-29.

[15] Nilsson N.(1982) "Principles of Artificial Intelligence", *Springer:Berlin.*

[16] Tsay R.S. and G.C. Tiao(1984) "Consistent estimates of autoregressive parameters and extended sample autocorrelation function for stationary and nonstationary ARMA models", *JASA*, 79, 84-96.

[17] Weitz R.R.(1986) "NOSTRADAMUS A knowledge-based forecasting advisor", *Int.J.Forecast.*, 2, 273-283.

DOMINANCE AND INDIFFERENCE RELATIONS ON FUZZY NUMBERS

A. GONZALEZ and M.A. VILA
Dpto. de Ciencias de la Computación e I.A.
Facultad de Ciencias, Universidad de Granada,
18071-Granada (Spain).

ABSTRACT. A theoretical method for ranking fuzzy numbers is presented. It has good general properties and allow us to define different dominance relations. It leads to an improved indifference relation, in comparison with previous models, by means of a generalization of the ranking function approach that replaces $\mathbb{R}$ by $\mathbb{R}^m$. We study some properties of the ranking approach, and by interpreting the different parameters used to define it, we check that is capable of adaptation to the decision-maker's preferences. Finally, we have selected some difficult examples to see the behaviour of our ranking method.

1. Introduction

Fuzzy numbers were introduced in order to model imprecise situation involving real numbers. Comparison of fuzzy numbers have been treatted by several authors (Delgado et al. [5], Dubois and Prade [6],...); one of the more interesting and intuitive approach is the ranking function (Campos and González [3], Yager [15],...). This method maps fuzzy numbers into real numbers in such manner that the resulting numbers give us a meaningful way for ordering the original fuzzy numbers.

In this work we study several interesting properties of an order relation on the fuzzy numbers set defined initially in [9] and [10]. This order relation was defined by means of a generalization of the ranking function approach that replaces $\mathbb{R}$ by $\mathbb{R}^m$. The ranking functions on $\mathbb{R}^m$ are interpreted as the selection of m positions measures on a fuzzy quantity. Actually, when m=1, all information about a fuzzy quantity concentrates on a real number, and this fact may give "strange" indifference cases. Therefore, unlike the other methods for ranking fuzzy numbers, our model provides a good indifference relation (in some case, it is a discrete weakening of the strict equality). Moreover, our approach allows to define different dominance relations on the fuzzy numbers set, through the selection of different order relations on $\mathbb{R}^m$. In particular, strong and lexicographical order, generate strong and weak dominance, respectively.

Finally, we prove that this ranking approach can be considered as a

W. H. Janko et al. (eds.), Progress in Fuzzy Sets and Systems, 75–89.

weakening in several conditions of the comparison generated by using the fuzzy extension of the ordinary max operator.

2. Preliminary concepts

In this work, according to Goguen's Fuzzification Principle, we will call all fuzzy sets of the real line a **Fuzzy Quantity**, denoting the set of them as $\tilde{P}(\mathbb{R})$.

A fuzzy number is a particular case of a fuzzy quantity with the following properties:

Definition 2.1

The fuzzy quantity A with membership function $\mu_A(\cdot)$ is a **Fuzzy Number** iff:

i) $\forall\alpha\in[0,1]$, $A_\alpha=\{x\in\mathbb{R}/\ \mu_A(x)\geq\alpha\}$ (α-cuts of A) is a convex set.

ii) $\mu_A(\cdot)$ is an upper semicontinuous function (u.s.c.).

iii) $Supp(A)=\{x\in\mathbb{R}/\ \mu_A(x)>0\}$ is a bounded set of $\mathbb{R}$. □

Hence, the fuzzy numbers are fuzzy quantities whose α-cuts are closed and bounded intervals

$$A_\alpha=[a_\alpha,b_\alpha],\ \text{with}\ \alpha\in(0,1].$$

When $\alpha=0$, we consider A_0 as the closure of supp(A), i.e., $A_0=\overline{supp(A)}$. In the following text, $\tilde{\mathbb{R}}$ will denote the set of fuzzy numbers.

3. Schema of general ranking

Different methods for ranking fuzzy numbers, have been developed in the current literature. One of the more interesting and intuitive approach is the well-known ranking method through a function and an ordered set.

Let T be a generic set and $(O,\leq)$ an ordered set. We suppose the function

$$f:T\longrightarrow O$$

to be defined.

If we consider the equivalence relation associated to f

$$\forall A,B\in T\ \ AR_fB \Leftrightarrow f(A)=f(B)$$

we have the following function on the quotient set:

$$\tilde{f}: T_f \longrightarrow O \qquad / \; T_f = T/R_f$$

given by $\tilde{f}([A])=f(A)$. Obviously, $\tilde{f}$ is well defined and it is injective. Moreover, the function $\tilde{f}$ leads to the following **order relation** on the set T_f

$$[A],[B]\in T_f, \quad [A]\le[B] \iff \tilde{f}([A])\le\tilde{f}([B]).$$

Obviously, if $(O,\le)$ is a totally ordered set, then $(T_f,\le)$ is a totally ordered set too.

An order relation on T cannot be directly defined from f. Nevertheless, the order relation defined on T_f is valid for ranking elements of T, when the elements of a class of T_f are very homogeneous.

We denote by $\simeq_f$ to the **indifference relation** (or $\simeq$ when the function f is known):

$$A\simeq_f B \iff A,B\in[A] \iff f(A)=f(B).$$

The indifference problem is equivalent to the lack of discrimination of an index, and the equivalence class of the relation R_f contains those elements for which the index f is unable to discriminate between.

In the following, we consider T as a fuzzy quantity subset, and f will be called a **Ranking function**.

Several authors have used this approach by taking $\mathbb{R}$ as the ordered set (Adamo [1], Chang [4], Yager [15],...); this selection has the following problem: A ranking function on $\mathbb{R}$ may be seen as the selection of a position parameter on $\mathbb{R}$ for a fuzzy quantity. Thus, all the information about a fuzzy quantity is concentrated on a real number, which may give rise to "strange" cases of indifference (e.g. see Bortolan and Degani [2]).

So, we consider the indifference as an important problem for any ranking method, acting in our opinion as a test of goodness. The greater the similarity between the fuzzy quantities it makes equivalent, the better the ranking function.

In this work we shall therefore use ranking functions on $\mathbb{R}^m$, with $m\in\mathbb{N}$. These are interpreted as the selection of m position parameters on a fuzzy quantity. In this case, equality between the ranking functions implies equality between m position parameters, thus improving the indifference relation.

4. Making a continuous problem discrete

From an analytical point of view, the basic problem in ranking fuzzy numbers is the evaluation on continuous sets of their membership functions, and therefore the necessity of ranking curves in $\mathbb{R}^2$. To facilitate the resolution of the problem, we can represent a fuzzy quantity A as the set of its α-cuts, $\{A_\alpha; \alpha\in(0,1]\}$, and consider this two elements for making the ranking function:

1. We select a finite set of the unity interval, $Y\subset[0,1]$, then, $\{A_\alpha; \alpha\in Y\}$ may be considered a discrete version of the fuzzy quantity, that we use to develop a comparison between fuzzy quantities. We call **Ranking System** to any finite subset of the unity interval.

2. We define for every α in the ranking system $Y=\{\alpha_1,\alpha_2,\ldots,\alpha_n\}$ k representative position parameters on $\mathbb{R}$ (usually k=1 or 2), for such α-cut of the fuzzy quantity A

   $$p(A_{\alpha_i})\in\mathbb{R}^k, \quad i=1\ldots n$$

   that is, we assume $p(A_{\alpha_i})$ as k real parameters to be defined on each α_i-cut of A, representing the position of the set of real numbers A_{α_i} on $\mathbb{R}$. For example, the position of the set made up of a single real number is represented by a single parameter, namely the number itself. The position of the set of real numbers formed by an interval can be defined by two position parameters, i.e. the ends of the interval, etc.

By using the former elements the following standard ranking function can then be given

$$f:\tilde{P}(\mathbb{R})\longrightarrow\mathbb{R}^m$$

$$f(A)=(p(A_{\alpha_1}),p(A_{\alpha_2}),\ldots,p(A_{\alpha_n}))\in\mathbb{R}^m \;/\; m=nk \text{ and } A\in\tilde{P}(\mathbb{R}).$$

By using this function, an order relation on $\mathbb{R}^m$, and the previous general ranking schema, a comparison on $\tilde{P}(\mathbb{R})$ can be made.

If we use the strong order $(\le_S)$ on $\mathbb{R}^m$, that is,

$\forall x,y\in\mathbb{R}^m$, $x=(x_1,\ldots,x_m)$, $y=(y_1,\ldots,y_m)$

$$x \le_S y \Leftrightarrow x_i \le y_i \quad \forall i=1\ldots m.$$

the comparison is equivalent to the domination of all the position parameters of the α-cuts, with the usual order on $\mathbb{R}$. It generates a **Strong Dominance** between fuzzy quantities:

$$\text{"B strong-dominates A"} \Leftrightarrow f(A) \le_S f(B).$$

With the lexicographical order ($\le_L$), that is,

$$x \le_L y \Leftrightarrow \begin{cases} \exists k \;/\; x_i = y_i \text{ with } 0<i<k \text{ and } x_k<y_k \\ \text{or} \\ x_i = y_i \quad \forall i=1\ldots m. \end{cases}$$

and based on a priority between the different α in the ranking system, the standard ranking function compares parameter to parameter, the first non-indifferent to be encountered leading to the solution of the comparison. It generates a **Weak Dominance** between fuzzy quantities:

$$\text{"B weak-dominates A"} \Leftrightarrow f(A) \le_L f(B).$$

Obviously, the existence of strong dominance between fuzzy quantities implies the existence of weak dominance as well.

When the nk position parameters coincide, indifference is obtained. Through the suitable selection of the position parameters, it is always possible to fit the relation R_f to an indifference admissible by the decision-maker. Hence it makes no difference whether $\tilde{P}(\mathbb{R})$ or the quotient set is ranked.

5. Particular ranking function on $\tilde{\mathbb{R}}$

5.1 DEFINITION OF f_T

Let $A \in \tilde{\mathbb{R}}$, and let $Y=\{\alpha_1, \alpha_2, \ldots, \alpha_n\}$ be a ranking system with $\alpha_i \in [0,1]$, such that $A_{\alpha_i} \ne \emptyset$, $\forall i \in I=\{1,2,\ldots n\}$. A is a convex set and μ_A is an u.s.c. function, therefore A_{α_i} is a closed interval:

$$A_{\alpha_i} = [a_i, b_i], \quad \forall i \in I.$$

To define a particular ranking function, two position parameters (k=2) representing two points in each α-cut are selected. They are chosen through two parameters λ and μ on the unity interval

$$p(A_{\alpha_i})=(\lambda b_i+(1-\lambda)a_i,\mu b_i+(1-\mu)a_i)$$

and we define the following ranking function.

Definition 5.1

We define the ranking function

$$f_T:\tilde{\mathbb{R}}\times[0,1]^2\longrightarrow\mathbb{R}^{2n}$$

$$f_T(A,\lambda,\mu)=(\lambda b_1+(1-\lambda)a_1,\mu b_1+(1-\mu)a_1,\ldots,\lambda b_n+(1-\lambda)a_n,\lambda b_n+(1-\lambda)a_n)$$

in short

$$f_T(A,\lambda,\mu)=(\lambda b_i+(1-\lambda)a_i,\mu b_i+(1-\mu)a_i)_1^n$$

Based on the parameter λ and μ, and by means of an order relation on $\mathbb{R}^{2n}$, the function f_T provides an order relation between elements of the quotient set $\tilde{\mathbb{R}}_{f_T}$.

5.2 INDIFFERENCE GENERATED BY f_T

The indifference is given by:

$$\forall A,B\in\tilde{\mathbb{R}},\ A_{\alpha_i}=[a_i,b_i],\ B_{\alpha_i}=[c_i,d_i]$$

$$A\simeq B \iff f_T(A,\lambda,\mu)=f_T(B,\lambda,\mu) \iff \begin{cases}\lambda b_i+(1-\lambda)a_i=\lambda d_i+(1-\lambda)c_i\\ \mu b_i+(1-\mu)a_i=\mu d_i+(1-\mu)c_i\end{cases} \forall i\in I$$

When the parameter λ and μ are different, the indifference relation is equivalent to the coincidence between the respective α-cuts of the ranking system on each fuzzy number, as can be easily proved.

Thus, the indifference generated by f_T is a discrete weakening of the usual equality, because of the equality between fuzzy sets defined by Zadeh $(\mu_A=\mu_B)$ is equivalent to the equality between their α-cuts respectively, and the indifference generated by the function f_T is equivalent to the equality between a finite number of α-cuts, that is,

$$A=B \iff A_\alpha=B_\alpha \quad \forall\alpha\in(0,1]$$
$$A\simeq B \iff A_\alpha=B_\alpha \quad \forall\alpha\in\{\alpha_1,\alpha_2,\ldots,\alpha_n\}.$$

By using as ranking system $Y=\{\alpha_1,\alpha_2,\alpha_3,\alpha_4\}$ the fuzzy numbers represented in figure 1 are indifferent.

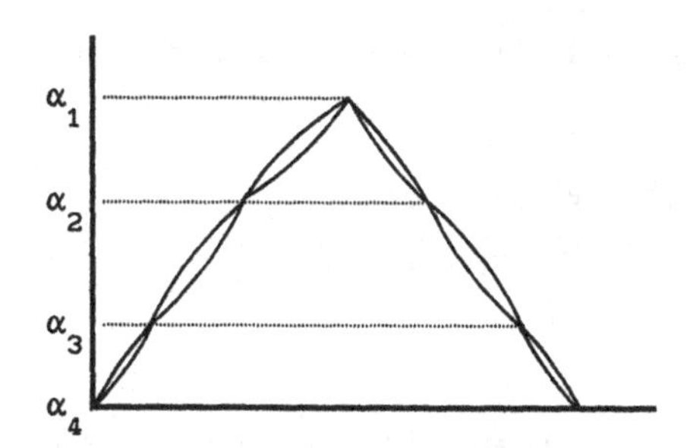

Figure 1. Two indifferent fuzzy numbers.

Obviously, this indifference relation improves the more elements in the ranking system we consider.

For a triangular fuzzy number two different elements in the ranking system are sufficient, in order for the relation $\simeq$ (with $\lambda \neq \mu$) to coincide with Zadeh's equality.

The ranking systems are an alternative to the continuous consideration of the fuzzy sets. In this way, the most inmediate selection of a ranking system consists in an "exploration" of the unity interval. For example,

$$Y_n = \{0 \cup i/n,\ i=1\ldots n\} \quad \text{with } n \in \mathbb{N},$$

can be good ranking systems.

6. General properties of f_T

Some important properties of the ranking function will be examined below.

Proposition 6.1

The order relation generated by f_T restricted to the real number set coincides with the usual order relation.

Proof.

By considering each real number as a singleton of greatest height, then

$$A \in \mathbb{R} \subset \tilde{\mathbb{R}} \Rightarrow A_\alpha = A \ \forall \alpha \in (0,1],$$

and

$$f_T(A, \lambda, \mu) = (A, A, \ldots, A) \in \mathbb{R}^{2n}$$

and any order relation considered on $\mathbb{R}^{2n}$ maintains the usual order.□

Proposition 6.2

Let $A,B\in\tilde{\mathbb{R}}$, $r\in\mathbb{R}$. We assume $\oplus$ to be the extended addition of fuzzy numbers. Then

i) $f_T(A\oplus B,\lambda,\mu)=f_T(A,\lambda,\mu)+f_T(B,\lambda,\mu)\quad \forall\lambda,\mu\in[0,1]$,

ii) $f_T(rA,\lambda,\mu)=rf_T(A,\lambda,\mu)\quad \forall\lambda,\mu\in[0,1]$.

Proof.

By using the following result (Nguyen [12] for α-cuts and Negoita [11] for strong α-cuts):

$$\forall A,B\in\tilde{\mathbb{R}}\quad (A\oplus B)_{\alpha_i}=A_{\alpha_i}+B_{\alpha_i}\quad\text{and}\quad (rA)_{\alpha_i}=rA_{\alpha_i}$$

for any ranking system $Y=\{\alpha_1,\alpha_2,\ldots,\alpha_n\}$, and for any pair of parameters λ and μ of the unity interval. Therefore the result is obvious.□

This result demonstrates the linearity of the ranking function on fuzzy numbers.

The following proposition was demonstrated in [8] and shows the compatibility between the order generated by f_T and the extended addition by one hand and the product of positive fuzzy numbers by other hand. A positive fuzzy number A is defined by $A_\alpha\subset\mathbb{R}^+$, $\forall\alpha\in\{\alpha_1,\alpha_2,\ldots,\alpha_n\}$. Dubois and Prade in [7] define a stronger definition of positive fuzzy number, by considering $A_\alpha\subset\mathbb{R}^+$, $\forall\alpha\in(0,1]$. We denote $\tilde{\mathbb{R}}^+$ the positive fuzzy numbers set.

Proposition 6.3

The following expressions hold for the strong or lexicographical order on $\mathbb{R}^{2n}$

i) $f_T(A,\lambda,\mu)\le_Z f_T(B,\lambda,\mu)$, $f_T(C,\lambda,\mu)\le_Z f_T(D,\lambda,\mu)\Rightarrow$

$$f_T(A\oplus C,\lambda,\mu)\le_Z f_T(B\oplus D,\lambda,\mu),\quad \forall A,B,C,D\in\tilde{\mathbb{R}},\ \forall\lambda,\mu\in[0,1].$$

ii) $f_T(A,\lambda,\mu)\le_Z f_T(B,\lambda,\mu)$, $f_T(C,\lambda,\mu)\le_Z f_T(D,\lambda,\mu)\Rightarrow$

$$f_T(A\odot C,\lambda,\mu)\le_Z f_T(B\odot D,\lambda,\mu),\quad \forall A,B,C,D\in\tilde{\mathbb{R}}^+,\ \lambda=0,\mu=1 \text{ or } \lambda=1,\mu=0.$$ □

We study now the behaviour of the ranking function by considering changes in the fuzzy number scale. This is an interesting problem when we work with fuzzy numbers obtained with different units of measure, or when we want to place all the variables into a new interval, such as for example the unity interval. The following result shows that the order and indifference on fuzzy numbers are independent of the scale used.

Proposition 6.4

Let $A,B\in\tilde{\mathbb{R}}$, $\lambda,\mu\in[0,1]$, and let $A^e,B^e\in\tilde{\mathbb{R}}$ be the transformed fuzzy numbers of A,B by a scale change $e(x)=rx+s$, with $r,s\in\mathbb{R}$, $r>0$. The following expressions then hold for the strong or lexicographical order on $\mathbb{R}^{2n}$,

i) $f_T(A,\lambda,\mu)\le f_T(B,\lambda,\mu) \Leftrightarrow f_T(A^e,\lambda,\mu)\le f_T(B^e,\lambda,\mu)$

ii) $A\simeq B \Leftrightarrow A^e\simeq B^e$.

Proof.

It is easily checked with the arithmetic of fuzzy quantities that

$$\mu_{A^e}=\mu_A\circ e \Leftrightarrow A^e=\frac{1}{r}(A\ominus s)$$

By using Proposition 6.2

$$\forall\lambda,\mu\in[0,1]\ \ f_T(A,\lambda,\mu)=rf_T(A^e,\lambda,\mu)+(s,\ldots,s),$$

and therefore the result is obvious.□

7. Dominance regions on fuzzy numbers

When the parameters λ and μ, in the ranking function, are equal, then we call to $f_T(\cdot,\lambda,\lambda)$ **Average Ranking Function**, and we note

$$\forall A\in\tilde{\mathbb{R}}\quad f_T(A,\lambda,\lambda)=f_\lambda(A).$$

The average function selects a single average point, to represent the position of every α-cut $A_\alpha=[a_\alpha,b_\alpha]$. The parameter λ can be interpreted as an optimism-pessimism degree, which must be selected by the decision-maker.

So, when the most advantageous decision is to choose the greatest quantity, an optimistic person would think at the upper extreme of the interval b_α ($\lambda=1$), reflecting the greatest possible profit. On the contrary, a pessimistic person would prefer the lower extreme of the interval a_α ($\lambda=0$), representing the least possible profit.

When the most advantageous decision is to choose the least quantity, the interpretation is the opposite, with $\lambda=0$ for optimism and $\lambda=1$ for pessimism. Between the two extreme values $\lambda=0$ and $\lambda=1$ there is an attitudes scale in front of the uncertainty for every decision-maker. When the decision-maker does not profit of the decision, then $\lambda=1/2$ could reflect this situation.

$\lambda=0$	$\lambda=0.5$	$\lambda=1$
a_α		b_α
the least possible profit		the greatest possible profit

Another interesting interpretation of the parameter λ is to consider it as a requirement level for the decision model.

If the parameter λ is not known "a priori", then it is always possible to calculate the **Dominance Region**

$$R_z(A,B)=\{\lambda\in[0,1] \;/\; f_\lambda(A)\leq_z f_\lambda(B)\}$$

with z=s or L (strong or lexicographical order). The computation of this region can be used in the orientation for ranking A and B. This one provides a previous overall view of the decision, and it makes easy to take an optimistic or pessimistic posture. Moreover, it allows to the decision-maker to know the sensitivity of the parameter λ, i.e., we can also know whether a little change of the parameter modifies or not the final decision.

We study now different properties of the dominance region. The following proposition shows that the dominance region is a convex set.

Lemma.

Let $A\in\tilde{\mathbb{R}}$, $\lambda,\mu\in[0,1]$. Then

$$f_{t\lambda+(1-t)\mu}(A)=tf_\lambda(A)+(1-t)f_\mu(A),\ \forall t\in[0,1].$$

Proof.

Let $Y=\{\alpha_1,\alpha_2,\ldots,\alpha_n\}$ a ranking system, $A_{\alpha_i}=[a_i,b_i]$,

then

$$f_{t\lambda+(1-t)\mu}(A)=((t\lambda+(1-t)\mu)b_i+(1-t\lambda-(1-t)\mu)a_i)_1^{2n}=$$

$$=(t(\lambda b_i+(1-\lambda)a_i)+(1-t)(\mu b_i+(1-\mu)a_i)_1^{2n}=tf_\lambda(A)+(1-t)f_\mu(A).\square$$

Proposition 7.1

Let $A,B\in\tilde{\mathbb{R}}$, then $R_z(A,B)$ is either $\emptyset$ or is an interval contained in $[0,1]$, with z=s or L.

Proof.

We suppose $R_z(A,B)\neq\emptyset$. Let $\lambda,\mu\in R_z(A,B)$. By definition of dominance region, we hold

$$f_\lambda(A)\le_z f_\lambda(B) \text{ and } f_\mu(A)\le_z f_\mu(B),$$

therefore

$$tf_\lambda(A)+(1-t)f_\mu(A)\le_z tf_\lambda(B)+(1-t)f_\mu(B) \quad \text{with } t\in[0,1].$$

Applying the previous lemma

$$f_{t\lambda+(1-t)\mu}(A)\le_z f_{t\lambda+(1-t)\mu}(B) \quad \text{and} \quad t\lambda+(1-t)\mu\in R_z(A,B),$$

therefore $R_z(A,B)$ is a convex set of $\mathbb{R}$ and an interval.□

We study now the set

$$I(A,B)=R_z(A,B)\cap R_z(B,A)$$

which contains the parameters where the fuzzy numbers A and B are indifferent. This region is independent of the order relation z.

Proposition 7.2

Let $A,B\in\tilde{\mathbb{R}}$, then $I(A,B)$ is either $\emptyset$ or $[0,1]$ or a single point.

Proof.

We consider the equality between the value of f_λ of A and B:

$$f_\lambda(A)=f_\lambda(B) \iff \lambda f_1(A)+(1-\lambda)f_0(A)=\lambda f_1(B)+(1-\lambda)f_0(B) \iff$$

$$\iff \lambda(f_1(A)-f_0(A)-f_1(B)+f_0(B))=f_0(B)-f_0(A).$$

$$f_0(B)-f_0(A)=(x_i)_1^{2n}\in\mathbb{R}^{2n}, \qquad f_1(B)-f_1(A)=(y_i)_1^{2n}\in\mathbb{R}^{2n}.$$

So,

$$\lambda(x_i-y_i)_1^{2n}=(x_i)_1^{2n} \Rightarrow \lambda(x_i-y_i)=x_i \quad \forall i$$

then

if $x_i=0\ \forall i$ and $x_i=y_i\ \forall i \Rightarrow I(A,B)=[0,1]$.

if $x_i=0\ \forall i$ and $\exists j\ x_j\neq y_j \Rightarrow I(A,B)=\{0\}$.

if $\exists j\ x_j\neq 0$ and $x_i=y_i\ \forall i \Rightarrow I(A,B)=\emptyset$.

if $\exists j\ x_j\neq 0$ and $\exists k\ x_k\neq y_k \Rightarrow$

$$\Rightarrow \begin{cases} \text{if } \dfrac{x_1}{x_1-y_1} = \dfrac{x_2}{x_2-y_2} = \ldots = \dfrac{x_n}{x_n-y_n} = \lambda \Rightarrow I(A,B)=\{\lambda\} \\ \text{otherwise} \Rightarrow I(A,B)=\emptyset \end{cases} . \square$$

The region

$$C_Z(A,B)=R_Z(A,B)\cup R_Z(B,A)$$

contains the parameters such that A and B are comparable. Obviously, $C_L(A,B)=[0,1]$, but in general

$$N_S(A,B)=[0,1]-C_S(A,B)$$

contains to the parameter where A and B are not strictly comparable.

Proposition 7.3

Let $A,B\in\tilde{\mathbb{R}}$, then $R_S(A,B)\subseteq R_L(A,B)$

Proof.

Since the strong dominance implies the weak dominance, this result is straightforward. $\square$

The following proposition shows how we use the dominance region in the general ranking function and the strong order, and how μ is a reserve parameter of optimism-pessimism to guarantee a good level of indifference.

Proposition 7.4

Let $A,B\in\tilde{\mathbb{R}}$, $\lambda,\mu\in[0,1]$, then

$$f_T(A,\lambda,\mu)\le_S f_T(B,\lambda,\mu) \Leftrightarrow \lambda,\mu\in R_S(A,B) \Leftrightarrow [\lambda,\mu]\subseteq R_S(A,B).$$

Proof.

Let $A_{\alpha_i}=[a_i,b_i]$ and $B_{\alpha_i}=[c_i,d_i]$, then

$$f_T(A,\lambda,\mu)\le_S f_T(B,\lambda,\mu) \Leftrightarrow \begin{cases} \lambda b_i+(1-\lambda)a_i\le\lambda d_i+(1-\lambda)c_i \\ \mu b_i+(1-\mu)a_i\le\mu d_i+(1-\mu)c_i \end{cases} \forall i \Leftrightarrow$$

$$\Leftrightarrow f_T(A,\lambda,\lambda)\le_S f_T(B,\lambda,\lambda) \text{ and } f_T(A,\mu,\mu)\le_S f_T(B,\mu,\mu) \Leftrightarrow$$

$$\Leftrightarrow \lambda,\mu\in R_S(A,B) \Leftrightarrow [\lambda,\mu]\subseteq R_S(A,B).$$

this last equivalence is obtained by using that $R_S(A,B)$ is a convex set. □

By using proposition 7.4 we can check that $\lambda=0$ and $\mu=1$ are the strongest possible choice of these parameters, since

$$\text{if } 1,0 \in R_S(A,B) \text{ then } R_S(A,B)=[0,1]$$

and $f_T(A,\lambda,\mu) \leq_S f_T(B,\lambda,\mu)$ $\forall \lambda,\mu \in [0,1]$.

Finally, by considering the strongest conditions of the ranking approach,

- $Y=[0,1]$.
- Strong order on $\mathbb{R}^m$.
- $\lambda=0, \mu=1$.

we prove that the ranking method generated by f_T with these conditions is equivalent to use the fuzzy extension of the ordinary max operator..

Proposition 7.5

Let $Y=[0,1]$, $\lambda=0, \mu=1$, then $A \leq_S B \Leftrightarrow \underset{\sim}{\max}\{A,B\}=B$.

Proof.

Let $A_\alpha=[a_\alpha,b_\alpha]$ and $B_\alpha=[c_\alpha,d_\alpha]$, then

$$A \leq_S B \Leftrightarrow f_T(A,0,1) \leq_S f_T(B,0,1) \Leftrightarrow \begin{cases} a_\alpha \leq c_\alpha \\ \\ b_\alpha \leq d_\alpha \end{cases} \quad \forall \alpha \in [0,1] \Leftrightarrow \underset{\sim}{\max}\{A,B\}=B. \square$$

Therefore this ranking approach can be considered as a weakening in all possible conditions ($Y \subseteq [0,1]$, $\lambda,\mu \in [0,1]$, $\leq_z$ with z any order relation on $\mathbb{R}^{2n}$) of the operator $\underset{\sim}{\max}$.

8. Examples

Finally, we show different examples representative of our ranking method. We are selected some of the most difficult examples among the large collection of cases proposed by Bortolan and Degani [2].

a) We consider the triangular fuzzy numbers

$$A=\{(2,4,1,1),1\},\ B=\{(4,4,3,1),1\},\ C=\{(4,4,1,1),1\}$$

represented in the fig.2a.

We consider the "exploratory" ranking system:

$$Y=\{0,0.25,0.50,0.75,1\}$$

and as ordered ranking system

$$Y_o=\{0.50, 0.75, 0.25, 1, 0\}.$$

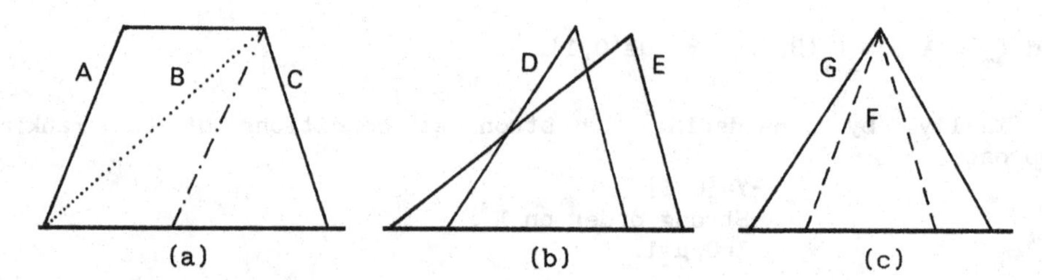

Figure 2. Some triangular fuzzy numbers.

The solution is

"C strong-dominates B" and "B strong-dominates A", $\forall\lambda\in[0,1]$
"A indifferent B" and "B indifferent C", with $\lambda=1$.

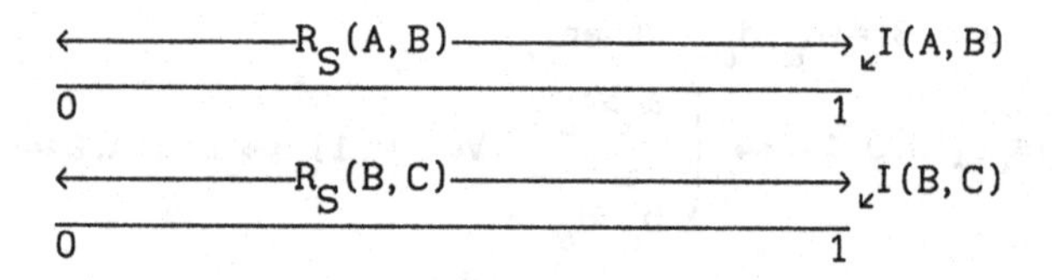

b) We consider the triangular fuzzy numbers

$$D=\{(2,2,2,1),1\},\quad E=\{(3,3,4,1),1\}$$

represented in the fig. 2b.
The result is

"E strong-dominates D" $\forall\lambda\in[0.5,1]$ (optimistic choices)
"E weak-dominates D" $\forall\lambda\in[0,1]$.

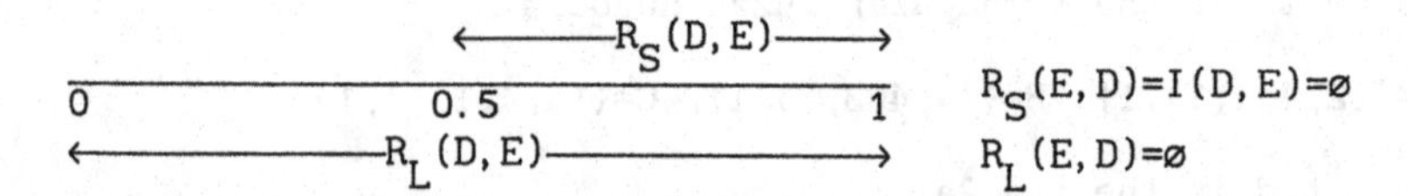

c) We consider the triangular fuzzy numbers

$$F=\{(4,4,1,1),1\},\quad G=\{(4,4,2,2),1\}$$

represented in figure 2c.

The classification is

An optimistic person would prefer G ($\lambda\in[0.5,1]$).
A pessimistic person would prefer F ($\lambda\in[0,0.5]$).
In an intermediate situation G and F are indifferent (λ=0.5).

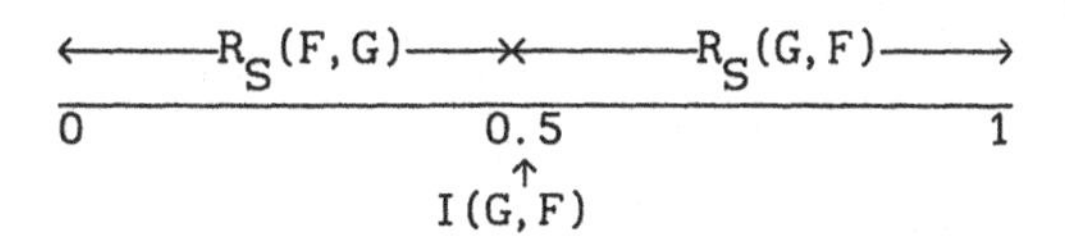

REFERENCES

[1] Adamo, J.M. (1980) 'Fuzzy decision trees'. Fuzzy Sets and Systems 4, 207-219.
[2] Bortolan, G. and Degani, P. (1985) 'A review of some methods for ranking fuzzy subsets' Fuzzy Sets and Systems 15, 1-19.
[3] Campos, L.M. de and González, A. (1989) 'A subjective approach for ranking fuzzy numbers' Fuzzy sets and system 29, n.1, 145-153.
[4] Chang, W. (1981) 'Ranking of fuzzy utilities with triangular membership function' Proc. Int. Conf. on Policy Anal. and Inf. Systems, 263-272.
[5] Delgado, M., Verdegay, J.L. and Vila, M.A. (1988) 'A procedure for ranking fuzzy number using fuzzy relations' Fuzzy Sets and Systems, 26, 49-62.
[6] Dubois, D. and Prade H. (1983) 'Ranking fuzzy numbers in the setting of possibility theory' Information Sciences 30, 183-224.
[7] Dubois, D. and Prade H. (1985) 'Fuzzy numbers. An overwiew' The Analysis of Fuzzy Information, in J.C. Bezdek (eds.) CRS Press, Boca Ratón F1,USA.
[8] González, A. (1987) 'Métodos subjetivos para la comparación de números difusos' Ph. D. Thesis, Universidad de Granada.
[9] González, A. and Vila, M.A. 'A discrete method to study indifference and order relations between fuzzy numbers' To appear in Information Science.
[10] González, A. and Vila, M.A. 'Dominance relations on fuzzy numbers' submitted to Information Science.
[11] Negoita, C.V. (1978) 'Management applications of systems theory' Birkhaüser Verlag, Basel.
[12] Nguyen, H.T.(1978) 'A note on the extension principle for fuzzy sets' Journal of Math. Anal. and Appl. 64, 369-380.
[13] Rámik, J. and Římánek, J. (1985) 'Inequality relation between fuzzy numbers and its use in fuzzy optimization' Fuzzy Sets and Systems, 16, 123-138.
[14] Tanaka, H., Ichihashi, H. and Asai, K. (1984) 'A formulation of fuzzy linear programming problem based on comparison of fuzzy numbers. Control and Cybernetics, vol.13, n.3.
[15] Yager, R.R. (1978) 'Ranking fuzzy subsets over the unit interval' Proc.1978 CDC 1435-1437.

A FUZZY CLASSIFICATION OF COMPUTER USERS ATTITUDES: A CASE STUDY

Dr M. KALLALA & Dr W. BELLIN
Department of Psychology
University of Reading
Whiteknights
Reading RG6 2AL
England

ABSTRACT. We present in this paper a brief outline of a general methodology that classifies computer users attitudes by fuzzifying information elicited in the course of human interaction with some system or machine. We illustrate how the methodology can combine different kinds of emphasis in Human-Computer Interaction research by analysing and discussing a concrete example: the attitudes of a subject towards a set of activities.

1. Introduction

The goal in human computer interaction research is to find the best way of bringing together the efforts of humans and the resources of computer systems on any given set of tasks. Recent trends in human computer interaction research place a heavy emphasis on the way users experience systems [1]. So it is an advantage , for a methodology if it can be applied with generality across the range of issues creating uncertainty about system and user characteristics. In general, our methodology is concerned with classifying and optimizing information about categories of users, tasks and software applications. That is to optimize functions relating user performance and machine performance in order to serve as a knowledge-based system for the design of software or for the study of the ergonomics of hardware. However, we think it is essential to investigate, as well, the psychology of the computer user using a computer in relation to other situations involving other activities and other persons.

We follow the paradigm known as "user-centred design" where the emphasis on psychological tests becomes greater since systems are developed by studying the behaviour of users and then modifying the system itself to increase the compatibility between the task requirements and the system's expectations of the user. However, we retain an affinity with more analytical approaches since we acknowledge the need for an appropriate formalization. The particular formalization adopted is an interpretation of fuzzy logic, a logic incorporating 'real-world' facts, see Zadeh [2], as each set of entities interacts with the other in a vague and uncertain manner.

The methodology to be presented is crucial for the middle phase of such an interview. Psychological interview techniques can give structure to the middle phase, so that the user's experience is reflected in his or her own terms with exploration of the underlying dimensionality of those terms. The example to be discussed occurred during text processing tasks which took place as part of normal activities, and hence required comparison between self perceptions when processing text with the system, and handling text in alternative ways.

W. H. Janko et al. (eds.), Progress in Fuzzy Sets and Systems, 90–99.

Briefly, the methodology consists of four steps:

Step-1: preprocessing;
Step-2: forming basic relations from interview data;
Step-3: forming new composed relations, from the relations obtained in Step-2, by means of fuzzy operators;
Step-4: examining and optimising results and Hasse diagrams obtained in Step-2 and Step-3.

We illustrate our theoretical claim by analysing and discussing the attitudes of one subject towards a set of activities.

2. Basic Concepts

Software designers must have a comprehensive understanding of the users' background and environment. This can be achieved by taking the following steps:

(i) studying users' cognitive, cultural and behavioural characteristics and their performance, see [3];
(ii) interviewing users and measuring their perception of the system and attitudes to what they are doing as well as performance;
(iii) inputting the measures obtained into a set of fuzzy classification procedures based on fuzzy transitional relations;
(iv) interpretation of the results so as to produce guidelines for the design of systems which more closely match the requirements and the capabilities of the users.

Hence, in this conceptualization, users' attitudes to performing tasks with systems can be examined by interviewing them about their perception of themselves and key people performing the tasks with and without the aid of the system, then giving the elicited information a mathematical formulation.

2.1. INTERVIEWING METHODS

Amongst advocates of user-centred design, there is a move towards basing design decisions on "contextual" interviews with people in their normal work environment [4]. A problem with this move concerns treatment of attitude data, since there is a tension between the demand for performance data , which seems more susceptible to rigorous treatment, and the requirement for realistic attitude assessment.

It is a characteristic of our methodology that it avoids any such tension since a focus on behaviour and a focus on attitudes can be handled as special cases of the same general approach. Extended discussion of interview results from a sample individual user of a system can illustrate this characteristic.

In a "contextual" interview as described by Whiteside et al. [4], there are three parts. The opening partt "explores the nature of the user's work and ... initial responses to the system". Then there is a middle part where users are observed and questioned while describing what they are doing and experiencing. Finally , the interpretation in terms of functionality is made mainly by the investigator in a way that will relate to the design iterations. In the middle phase the user is treated as a participant in the whole design process rather than merely an object of study.

2.2. MATHEMATICAL FORMULATIONS

In the general conceptualisation, we have entities such as users and systems with possible attributes: categories of users, benchmark tests, fuzzy measures and so on.

In contextual interviews where attitudes are examined, mathematically, users constitute the set

$$\text{users} = u_1, u_2, u_3, \ldots, u_n$$

and each user is presented with a set of situations relevant to human-machine interaction, constituting the set of elements

$$\text{elements} = e_1, e_2, e_3, \ldots, e_m$$

which describe the following situations:

e_1: myself using the computer
e_2: myself at a laboratory task not involving computers
e_3: myself using a typewriter
e_4: myself writing a letter using pen and paper
e_5: someone I know who is familiar with computers
e_6: myself working on an academic paper
e_7: my mother using a computer
e_8: my father using a computer
e_9: a non-American overseas student using a computer
e_{10}: an American student using a computer.

A panoply of fuzzy descriptors, representing the user's individual judgement of the elements, is elicited. These judgments are ranked and then the ranks are normalised in the closed interval [0,1]. These rankings represent a fuzzy continum and are used as a scale in the measurement of the applicability of the descriptors, thus forming fuzzy relations. To take a particular example, where the task of editing text was being used to evaluate bilingual Arabic/English editors, the set of elements was contextualized by reference to the task situation. So the element e_2 was meant to be the user before gaining computer experience.

Nine triads made up from the ten elements were discussed to ensure that the user provided his own set of descriptors which distinguished one member of a triad from the other two. The element e_1 recurred in each triad in order to retain relevance to text editing. The following set of descriptors was obtained:

d_1: individual working
d_2: strategic knowledge
d_3: correction easy
d_4: clarifying
d_5: no idea
d_6: in harmony

d_7: American standard
d_8: no idea
d_9: absorbing

There was no constraint to avoid recurrence of descriptors.

It is common to attempt in such an interview to make the user provide contrasts for individual descriptors, as if it were possible to construe the task in terms of simple polar opposites, such as ease of correction as opposed to difficulty in correction. (See, for example, Easterby-Smith [5]). This constraint on the interview was avoided on the grounds of maintaining relevance and realism. Kallala and Bellin [4] discuss the lack of necessity for such constraints in a general methodology.

2.3. FUZZY TRANSITIONAL RELATIONS

To study the pattern of attitudes to performing the task with the system, all nine elements were ranked for closeness or applicability of each descriptor. The rankings are normalised to give values in the closed interval [0,1], representing a fuzzy continuum and a scale in the measurement of applicability. Fuzzy relations are then formed from the normalized rankings.

Thus, given the set of elements and the set of descriptors, we have the following set of relations:

$$De: \text{elements} \times \text{descriptors},$$

from which can be derived the transitional relations

(i) element - to - element
(ii) descriptor - to - descriptor.

These to some degree, in the range [0,1], may hold between each of the two entities. The reason why the fuzzy relations of interest are transitional is because they are the result of binary operations which depend on the type of implication operator. In the case of (i) we have

$$\text{element} \rightarrow \text{descriptor} \rightarrow \text{element}$$

and for (ii) we have

$$\text{descriptor} \rightarrow \text{element} \rightarrow \text{descriptor}$$

Several surveys of the definitions of the implication operator can be found in the literature (see among others. Bouchon and Desprès [6], Bandler & Kohout [7] and Willmott [8]); a selection of these and the possible formulation of the fuzzy transitional relations are amply discussed in Kallala [9]. However, for completeness, we list in Table 2.1 those we used in this study.

Code	Operator	Degree of Implication
1	Standard sharp	1 if a < 1 or b = 1 0 if a = 1 and b < 1
2	Standard strict	1 if a <= b 0 if a > b
3	Standard star	1 if a <= b b if a > b
4	Gaines	min (1, b/a)
4'	Modified Gaines	min (1, b/a, (1-a)/(1-b))
5	Lukasiewicz	min (1, 1-a+b)
5.5	Kleene-Dienes Lukasiewicz	1 - a + ab
6	Kleene-Dienes	max (1-a, b)
7	Early Zadeh	max (1-a, min (a, b))
8	Willmott	min (max (1-a, b), max (1-b, a), min (1-a, b)))

Table 2.1 List of the definitions of the implication operators used in this study. For any a and b in [0,1] such that,

$$\min(a, b) = a \wedge b,$$

$$\max(a, b) = a \vee b.$$

Having obtained a normalized set of rankings from each user, our next step is to formulate fuzzy transitional relations based on the set, De, defined above. We then examine the semantics of the user-machine interaction in the light of the taxonomy depicted by Hasse diagrams which were obtained in a similar fashion to that described by Bandler & Kohout [10].

Among the degrees of implications we are especially interested in the followings:

* the degree to which element i evokes a descriptor which is also evoked by element j
* the degree to which descriptor i is preferred by an element (user) to descriptor j.

In Figure 2.1, an example of such a classification is depicted. It shows the classification of element-to-element compositions for a selection of operators at a half-upper cut. Table 2.2 summarizes the properties of the element-to-element relations for the sample user. A thorough examination of the other operators is presented elsewhere [11 & 12].

Operator / Properties	4	4'	5	5.5
Reflexive	√	√	√	√
Antisymmetric	√	√		
Transitive	√	√	√	√
Local Reflexive	√	√	√	√
Preorder	√	√	√	√
Order	√	√		

Table 2.2 This is a summary of the properties of the element-to-element relations of one sample user obtained with the operators Gaines, modified Gaines, Lukasiewicz and Kleene-Dienes Lukasiewicz cut at half-upper.

3 Data Interpretation

When subjects particularize the elements in the procedure, in terms of their own context of experience with a system, interviewer notes about the elements are needed to supplement the Hasse diagrams and information about the properties of the relations. Even so, there is much that can be interpreted in the diagrams themselves.

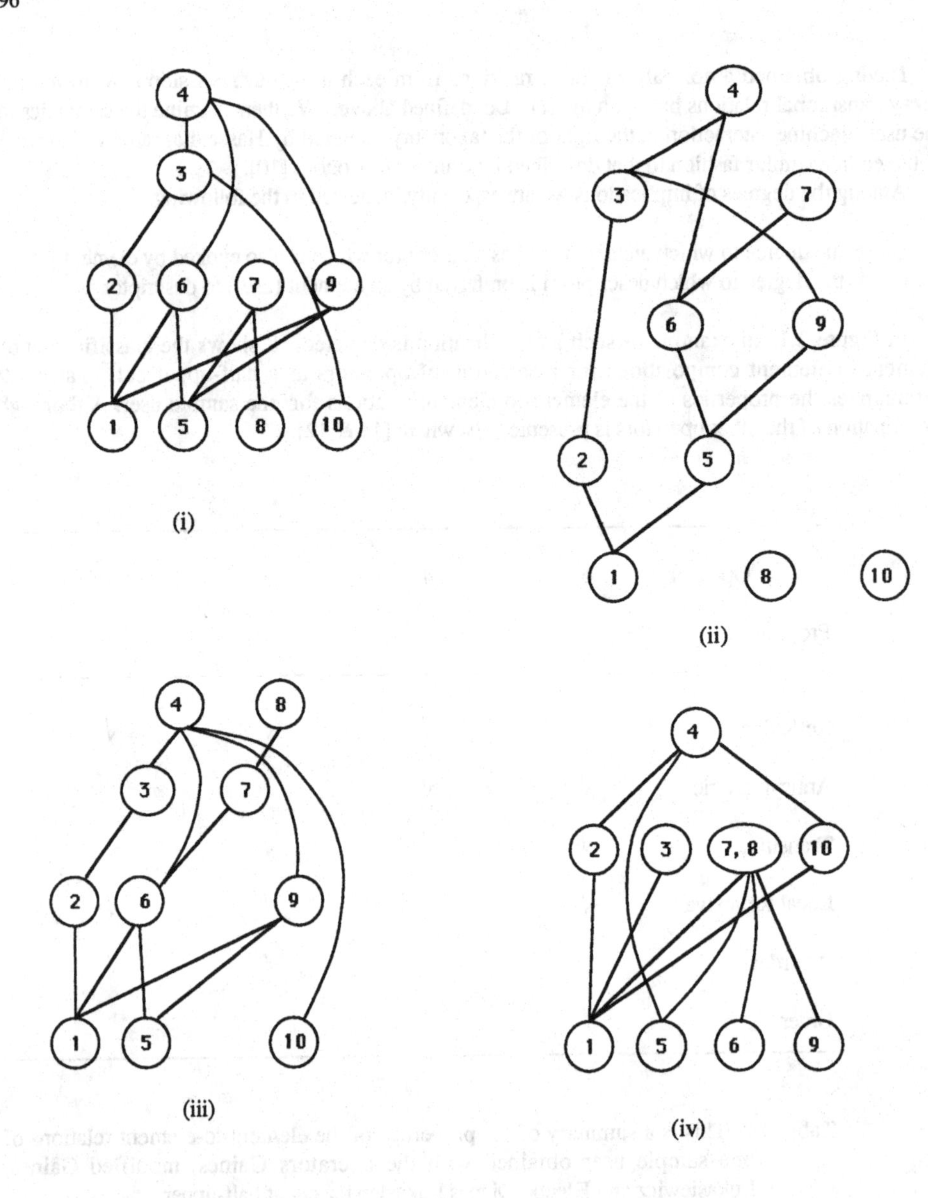

Fig. 2.1 Hasse diagrams showing the classification of the elements using the operators:

(i) 4: Gaines
(ii) 4': Modified Gaines
(iii) 5: Lukasiewicz
(iv) 5.5: Kliene-Dienes Lukasiewicz,

cut at half-upper.

3.1 CULTURAL AND EDUCATIONAL DIFFERENCES

In each Hasse diagram, there is consistently a wide distance between e_1 and e_4. In the context of using text editors, this can be interpreted as viewing a considerable discrepancy between working with text on the system, and working with text in other ways.

Across the diagrams, there is a recurrence of a relationship between e_4, e_3 and e_1 and e_5. Then there is some variation in the position of e_2 relative to e_{10}. What seems to be happening here is the reflection of an overall enthusiasm for information technology, and the descriptor "American standard" suggests a connection with symbolic values. So using the editor for text is connected with being very up to date, beyond the level of typewriter use or previous technical experience in laboratories. Nevertheless this surface enthusiasm needs to be weighted against the consistent wide distance between performing the tasks with pen and paper and using the system. Relevance to editing tasks is crucial for the functionality of the system, rather than general enthusiasm for being innovative.

Not all results can be interpreted clearly from the diagrams. For instance, in two of the Hasse diagrams e_8 is at a low level, but in two others it is at a high level. Table 2.2 is important here, since there is no order relation in the two diagrams where e_8 is high. Interview notes recorded ambivalence about comparisons of the father with other elements, since the user was not sure how lack of technical background as compared with other people among the elements might affect working with text using the editor.

3.2 SYSTEM FUNCTIONALITY

Results from contextual interviews are best interpreted alongside interview data with other people (see Kallala and Bellin [3]), and performance data. However, there was a very consistent feature of the results from this interview which alerted the investigators to a sensitive HCI issue.

"Myself writing a letter using pen and paper" (which was habitually in Arabic) was consistently distant from "an American student using a computer" and "myself using the computer" .Since there was a general overall favourability to adopting technology much as the "American" student, everyday tasks done with the system were seen as something well apart from performing the tasks in other ways. This did not happen with other users of the same system who did not work with the Arabic mode (see Kallala and Bellin [3]). The interview suggested that there is very low task compatibility between arabic text editing on this system and use of pen and paper.

The particular system being used had a low resolution character cell video display for Arabic characters based on a 14 by 9 array. Such resolution is regarded as adequate for discriminability of characters in text, and context sensitivity is handled by varying the amount of the cell for a character which is actually used. However, everyday working with text calls for more than adequate character discrimination. There needs to be overall task compatibility between working with a system and working in other ways.

Where editors with graphic interfaces are directly compared with character cell screen editors, it appears that Arabic speaking users strongly prefer superior graphics. Simple preference data can be dismissed by engineers on the grounds that they concern aesthetic preferences only. Contextual interviews suggest that the preference is more than aesthetic. For the experienced user,whose usage of the system covers a wide range of activities, making do with low resolution character cell interfaces does not seem too much of a compromise. This is because priorities in what such a user wants from a system are not mainly to do with text.

With much less experienced users, any overall favourability to using a computer system with text must not be taken at face value. To an Arabic speaker who has ever confronted an Arabic typewriter keyboard, the use of a computer keyboard is a tremendous advantage, since context sensitivity means there need be only one key for any Arabic orthographical symbol rather than separate keys (as on a typewriter) for every variant of the letter shapes.

However with character cell interfaces, resolution means there is a large difference between the appearance of a screen and a written or printed page. So a tolerance of discrepancy between computer use and text handling by other means hides a basic incompatibility between performing tasks on systems and using other means. Any overall favourability might be masking an appreciation of benefits from keyboarding with a simultaneous discomfort because of the quality of the visual interface. The situation is rather like the dissatisfaction that developed with nine pin dot matrix printers for Latin characters in spite of their widespread adoption. After a phase of adoption, users became more uncomfortable about the quality of Latin characters, and eventually turned to media that would give a more authentic quality.

4 Conclusion

We described here a methodology which is sufficiently general to cope with moves away from concentration on performance data in HCI towards assessing user attitudes to the tasks they are performing in contextual interviews. In the particular example discussed, a user compared himself at tasks with the system with himself and others doing the same tasks in alternative ways. It would have been just as easy to make the comparison between the self performing the task with the system as it was, and performing the same task with a prototype or augmentations of the system according to new specifications. In addition to generality, the methodology has practical applicability in deriving guidelines for software design wherever there is uncertainty about relevant user characteristics.

References

[1] Kallala, M. and Bellin, W. (1987) 'Categorisation of knowledge for handling uncertainties in the design of arabic software', in Rose, J. (ed.) Cybernetics and Systems: Present and Future, Proceedings of the Seventh International Congress of Cybernetics and Systems, September 7-11, 1987, London, England, Thales Press, Lytham St Annes, UK, pp. 757-761.

[2] Zadeh, L.A. (1975) 'The concept of linguistic variable and its application to approximate reasoning', Information Sciences 8, part I: 199-249 and part II: 301-357.

[3] Kallala, M. & Bellin, W. (1988) 'A study of arab computer users: a special case of a general hci methodology', in Bouchon, B., Saitta L. and Yager, R. R. (eds.), Uncertainty and Intelligent Systems, Lecture Notes in Computer Science 313, Springer-Verlag, pp. 338-350.

[4] Whiteside, J., Bennett, J., Holtzblatt, K. (1988) 'Usability engineering: our experience and evolution', in M. Helander. (ed.), Handbook of Human-Computer Interaction, North-Holland Press, Amsterdam, pp. 1-35.

[5] Easterby-Smith, M. (1980) 'The design, analysis and interpretation of repertory grids', Int. J. Man-Machine Studies 13, 3-24.

[6] Bouchon, B. and Desprès, S. (1986) 'Propagation of uncertainties and inaccuracies in

knowledge-based system', in Bouchon, B. and Yager, R.R. (eds.), Uncertainty in Knowledge-Based Systems, Lecture Notes in Computer Science 286, Springer-Verlag, pp. 58-65.

[7] Bandler, W. & Kohout, L.J. (1980) 'Fuzzy power sets and fuzzy implication operators', Fuzzy Sets and Systems 4, 13-30.

[8] Willmott, R. (1980) 'Two fuzzier implication operators in the theory of fuzzy power sets', Fuzzy Sets and Systems 4, 31-36.

[9] Kallala, M. (1986) 'Computer-aided Assessment of Dexterity of Neurological Patients by Means of Fuzzy Relational Products', PhD. Thesis, Brunel University, UK.

[10] Bandler, W. & Kohout, L.J. (1980) 'Semantics of implication operators and fuzzy relational products, Int. J. Man-Machine Studies 12, 89-116.

[11] Kallala, M. (1988) 'A measure-theoretic method for selection of fuzzy connectives in inferential structures of knowledge-based systems', Moscow Conference on FUZZY SETS IN INFORMATICS, Moscow, USSR, September 20-23, in proceedings thereof.

[12] Kallala, M. (1989) 'Fuzzy measure analysis: an optimisation of fuzzy connectives in the classification of parkinsonian's signatures', First European Congress on System Science, Lausanne, Switzerland, October 3-6, in proceedings thereof.

DIGITAL CIRCUITS BASED ON ALGEBRAIC FUZZY OPERATIONS

L. T. KÓCZY
Dept. of Communication Electronics
Technical University of Budapest
Stoczek u. 2, Budapest H-1111
Hungary

K. HIROTA
Dept. of Instrument & Control Engineering
College of Engineering, Hosei University
Kajino-cho 3-7-2, Koganei-shi, Tokyo 184
Japan

ABSTRACT. Some very interesting research has been done recently in the field of fuzzy computers. A basic element of a fuzzy computer able to perform multistep inference and suitable for storing fuzzy information necessarily contains fuzzy flip-flops (F^3), an extension of the traditional flip-flop concept. There are various approaches to F^3 the two most important ones are founded on min-max operations and algebraic operations, respectively. Algebraic fuzzy flip-flops are discussed in somewhat more detail. It is an obvious idea to investigate how digital functional units can be extended to fuzzy if instead of binary gates and flip-flops the fuzzy circuit components are applied. Especially the problem of "furry numbers" (fuzzy extension of binary numbers) and arithmetical operations on furry numbers is discussed. Also many open problems are presented.

1. INTRODUCTION

Some very interesting research including real implementations has been done recently in the field of fuzzy computers [1],[2],[3]. A basic element of a fuzzy computer able to perform multistep inference and suitable for storing fuzzy information necessarily contains fuzzy flip-flops (F^3), an extension of the traditional flip-flop concept. Of course, according to the great variety of possible fuzzy logical connectors fulfilling the necessary properties of **t**-norms and **t**-conorms (**s**-norms, respectively) [4], there are many possibilities to extend the concept of crisp digital circuits. For general investigations of various norms and detailed description of the F^3 see [5],[6],[7].

F^3 as a logical abstraction extends the idea of traditional two valued *JK* flip-flop.

1.1. The Models of the max-min F^3

Let us review quickly the idea of the binary *JK* flip-flop. In the

W. H. Janko et al. (eds.), Progress in Fuzzy Sets and Systems, 100–114.

following truth-table we summarize the behaviour of such a flip-flop. J and K are the inputs, Q the internal state in time t while Q^* is the new internal state i.e. in time $(t+1)$.

TABLE 1. Truth table of the JK flip-flop

J	K	Q	Q^*
0	0	0	0
0	0	1	1
0	1	0	0
0	1	1	0
1	0	0	1
1	0	1	1
1	1	0	1
1	1	1	0

From this table Q^* can be expressed by the minimal disjunctive form:

$$Q^* = J\bar{Q} + \bar{K}Q \tag{1}$$

and the minimal conjunctive form:

$$Q^* = (J + Q)(\bar{K} + \bar{Q}) \tag{2}$$

Of course these two equations are equivalent in Boolean algebra. Next step is extending the above two equations to the fuzzy case. As these equations are **not equivalent** in fuzzy algebra the mathematical model of F^3 can be obtained in two essentially different ways:

$$Q_S^* = (J \vee Q) \wedge (1-K) \vee (1-Q) \tag{3}$$

and

$$Q_R^* = \{J \wedge (1-Q)\} \vee \{(1-K) \wedge Q\}, \tag{4}$$

where $\vee$ and $\wedge$ stand for max and min, respectively. (3) and (4) are named the fundamental equations of **set type** and **reset type** F^3. It is true that

$$Q_S^* \geq Q_R^* \tag{5}$$

holds for every J, K and Q. For further details of the set-type and reset type F^3 see [5],[6],[7].

By the combination of the two equations (3) and (4) it is possible to achieve symmetricity of the F^3 and to extend the traditional idea of flip-flop smoothly for $J, K, Q \in [0,1]$:

$$Q^* = \begin{cases} (J \vee Q) \wedge (1-K) \vee (1-Q) & \text{if } J \geq K \\ \{J \wedge (1-Q)\} \vee \{(1-K) \wedge Q\} & \text{if } J \leq K \end{cases} \tag{6}$$

This is the fundamental equation of the min-max type F^3.

1.2. The algebraic F^3

This paper builds up the investigations on the algebraic fuzzy connectives (a special case of the interactive or *I*-fuzzy axiomatic system [8],[9]) fulfilling the **t**- and **s**-norm criteria in a *strictly* monotonous way.[6], [7]). The paper is going to touch the problem of F^3, from the aspect of **algebraic connectives** and it treats some questions in connection with the representation of fuzzy digital numbers and operations.

Expressing (1) and (2) with the norms we obtain

$$Q_R(t+1) = (J \mathbf{t} Q^n) \mathbf{s} (K^n \mathbf{t} Q) \quad (7)$$

$$Q_S(t+1) = (J \mathbf{s} Q) \mathbf{t} (K^n \mathbf{s} Q^n) \quad (8)$$

These equations define **reset type** and **set type** F^3 in general, similarly as (3) and (4) for the max-min connectives.

Let us replace now the norms by the usual algebraic representations:

$$A \mathbf{s} B = A + B - AB$$

and

$$A \mathbf{t} B = AB.$$

So the **algebraic set type** and **reset type** flip-flops are defined by

$$Q_S^A = F_S = J+Q-JQ-JKQ-KQ^2+JKQ^2 \quad (7)$$

and

$$Q_R^A = F_R = J+Q-2JQ-KQ+JQ^2+JKQ-JKQ^2 \quad (8)$$

Both formulas are rather complicated and there is a serious difficulty in obtaining a symmetrical and smooth extension in this case. In (3) and (4) if $J = K$, $Q_S = Q_R$, so the surface defined by (6) is continuous. Let us examine now (7) and (8) supposing $J = K$.

The condition for $F_S(J=K) = F_R(J=K)$ is

$$J+Q-JQ-J^2Q-JQ^2+J^2Q^2 = J+Q-3JQ+JQ^2+J^2Q-J^2Q^2$$

or

$$1+JQ = Q+J$$

which is generally not fulfilled.

So it is obvious that another way must be found for extending the idea of F^3 for algebraic operations. In order to find a solution to the problem of smooth extension a systematic analytical buildup of the algebraic F^3 has been done. In order to achieve this solution it was necessary first to introduce some basic postulates.

The starting point to these postulates was that every reasonable extension of the traditional flip-flop must preserve the rules in Table 1 as marginal conditions, i.e. F^3 must be identical with the usual *JK* flip-flop if $J,K,Q \in \{0,1\}$.

Let us start with the well known truth table of the crisp JK flip-flop (cf. Table 1) which can be expressed in a compressed form:

TABLE 2. Compressed truth table of the JK flip-flop

J	K	Q^*
0	0	Q
0	1	0
1	0	1
1	1	$\bar{Q}$

We shall denote from now on Q^* simply by F. We also remember that $\bar{A}$ means d(1,A) according to the I-fuzzy axiomatic system, i.e. $\bar{A} = 1 - A$.

According to the four lines of Table 2 we have now four postulates:

P1: $F(0,0,Q) = Q$
P2: $F(0,1,Q) = 0$
P3: $F(1,0,Q) = 1$
P4: $F(1,1,Q) = 1-Q$.

Beside the above four postulates we shall add also a fifth one. This postulate will express the fact that with inputs having maximal **fuzzy entropy** the new state of the F^3 must have also maximal entropy. For fuzzy entropy see e.g. [10]. Fuzzy entropy is maximal (= 1) if $J = K = 0.5$. So should be also Q and $\bar{Q}$. This postulate is very important as it formulates the fact that from ambiguous information only ambiguous result can be calculated. So we consider it one of the crucial properties of the algebraic F^3:

P5: $F(0.5,0.5,Q) = 0.5$,

where Q is arbitrary, i.e. F does not depend on the previous value of Q.

In the expressions (7) and (8) in both cases

$$Q(t+1) = F(Q(t),J,K)$$

is quadratic in $Q = Q(t)$ and linear in both J and K. By substituting the 5 postulates into a general parametrized expression of this type, a linear equation system for the parameters is obtained which has a unique solution (see [11]).

$$F = (\alpha_0+\alpha_1J+\alpha_2K+\alpha_3JK) + (\beta_0+\beta_1J+\beta_2K+\beta_3JK)Q + (\gamma_0+\gamma_1J+\gamma_2K+\gamma_3JK)Q^2 \quad (9)$$

where α_i, β_i and γ_i are reals.

By substituting the postulates into the above general parametrized expression, the following linear equation system is obtained for the parameters:

P1: $\alpha_0+\beta_0 Q+\gamma_0 Q^2 = Q$

From here

$\alpha_0 = 0, \quad \beta_0 = 1, \quad \gamma_0 = 0$

P2: $\alpha_2+(1+\beta_2)Q+\gamma_2 Q^2 = 0$

$\alpha_2 = 0, \quad \beta_2 = -1, \quad \gamma_2 = 0$

P3: $\alpha_1+(1+\beta_1)Q+\gamma_1 Q^2 = 1$

$\alpha_1 = 1, \quad \beta_1 = -1, \quad \gamma_1 = 0$

P4: $1+\alpha_3)+(1-1-1+\beta_3)Q+\gamma_3 Q^2 = 1-Q$

$\alpha_3 = 0, \quad \beta_3 = 0, \quad \gamma_3 = 0$

As we said the equation system has a unique solution (although P5 was not at all used). This solution is:

$$F = J + (1-J-K)Q = J + Q - JQ - KQ \qquad (10)$$

Of course, it is not obvious that (10) defines really a fuzzy function F. It is easy to prove however that

$$F = J(1-Q) + Q(1-K)$$

which is an alternate form of (10), is always > 0 and < 1 if J, K, $Q \in (0,1)$.

It is easy to prove that P5 satisfies (10):

$$F(0.5,0.5,Q) = 0.5+Q-0.5Q-0.5Q = o.5$$

So (10) is solution and the only solution for P1...P5. It can be observed that (10) is quite simple in comparison to F_R and F_S. (10) will be named the **fundamental equation of the algebraic F^3.**

Some interesting properties of this algebraic F^3 can be proven. We give here only a short list of them and refer to further publications containing some more detailed discussions of this topic including also proofs [11],[12],[13]

Property 1: F is always > 0 if $J,K,Q \neq 0,1$

Property 2: F is always < 1 if $J,K,Q \neq 0,1$

Property 3: $F(J_1) > F(J_2)$ if $J_1 > J_2$ and $Q \neq 1$

It can be remarked that if $Q = 1$ $F \neq f(J)$, so $F(J_1) = F(J_2)$ for arbitrary J_1 and J_2.

Property 4: $F(K_1) < F(K_2)$ if $K_1 > K_2$ and $Q \neq 0$

Property 5: $F(D,1-D,Q) = D$

This property enables the introduction of D-F^3 which is a generalization of the usual D flip-flop concept.

Property 6: $F(T,T,Q) = 1-Q$ if $T = 1$ and Q if $T = 0$

This property enables similarly the extension of the idea of T flip-flop by the algebraic T-F^3:

$$F_T = T + (1-2T)Q = T + Q - 2TQ \tag{11}$$

Finally, we mention that (5) offers a possibility to generalize the F^3 for arbitrary connectives:

$$F = d(J \mathbf{s} Q, K \mathbf{t} Q) \tag{12}$$

$d(,)$ representing $| - |$, as e.g. $A^n = d(1,A)$ (see [8]).

2. FUZZY FUNCTIONAL UNITS AND FURRY NUMBERS

2.1. Registers and counters

The introduction of algebraic JK-F^3, D-F^3 and T-F^3 enables the generalization of various sequential circuit types as e.g. of counters, registers, etc. In this way the single F^3's will represent something being the extension of binary digits (bits) for fuzzy values. Fuzzy bits have been discussed by several authors and the use of the name **fit** was proposed [10]. In the case of a traditional (non-fuzzy) computer information is stored in binary flip-flops: every flip-flop storing one bit. Accordingly in fuzzy systems information is stored in F^3's in the form of fits. By constructing registers from F^3's the position of the fits becomes important: The significance of the fit positions is expressed by binary weights. It is an interesting open problem how to understand and how to use the fuzzy digital numbers obtained in this way. We proposed the calculation with such numbers first time in [14]. It is clear that the objects obtained by this method will be essentially different from the notion of fuzzy number which is more the extension of intervals as of single numbers.

2.2. Furry numbers

The fuzzy set obtained by a sequence of fits will be a set of discrete fuzzy numbers: a **fuzzy binary number** for which we propose the name **furry number**. This is an abbreviation of FUzzy binaRY number and at the same time it refers to the fact that elements having a membership degree different from 0 are located like the tips of hairs in the fur of some little furry animal. It is easy to construct shift registers from

F^3's used in *D*-flip-flop mode. Shifting to the left will exactly represent multiplication by 2 (maybe at the same time loosing the highest significance fit) and shifting to the left a division by 2 with rounding the result downward. When speaking about counters, a new problem will be raised: What is the equivalent of incrementation (count up) and decrementation (count down) for furry numbers? In the next section some ideas about arithmetics with furry numbers will be discussed.

A furry number is a vector of fits to the components of which a binary weight is assigned. So every fit is a fuzzy degree in $[0,1]$. In order to avoid confusion we are going to denote furry numbers by the sequence of their fits with an overbar. So e.g. $N = = \overline{0.2\ 0.6\ 0.8\ 0.3}$ is a four-fit number, where 2^0 has weight 0.3, 2^1 has 0.8, 2^2 is assigned 0.6, finally 2^3 has 0.2. Next question is: what is the meaning of N? Let us discuss first one fit. 0.3 means that the given fit is a fuzzy set of the bits 0 and 1 in such a way that $X_1 = \{0,1\}$ is the universe of discourse and the membership function expresses the degree in which the given fit is equal to 1. So

$$\mu(0) = 0.7,\ \mu(1) = 0.3$$

describes the lowest significanec fit. This concept can be extended to more than one bit and fit, respectively, in a similar way. N is a furry number consisting of 4 fits, i.e. it is a fuzzy set over the universe of discourse X_4, where X_4 is the set of all 4-bit binary numbers:

$$X_4 = \{0000,\ 0001,\ \ldots,\ 1110,\ 1111\}$$

Of course it brings no ambiguity if these numbers are represented in their decimal form:

$$X_4 = \{0,1,\ \ldots,\ 14,15\}$$

We can define N in such a way that it is representing e.g. $15_{10} = 1111_2$ in a degree $(0.2,0.6,0.8,0.3)$ which is a vectorial membership grade. Similarly $9_{10} = 1001_2$ is present in a degree of $(0.2,0.4,0.2,0.3)$, etc. When calculating a scalar degree we suggest to assign e.g. $0.2 \wedge 0.6 \wedge 0.8 \wedge 0.3$ to $\mu(15)$. $\wedge$ can be represented by various t-norms: if we adopt the min-max system we have $\mu(15) = 0.2$, while applying the algebraic product we obtain 0.0288.

<u>Statement 1</u>

Applying the algebraic product for $\wedge$

$$\sum_{X_i} \mu(x) = 1 \qquad (13)$$

for every $i \geq 1$.

Proof

Let us denote the jth fit by f_j, then

$$\mu(x) = \prod_{x_k=1} f_k \cdot \prod_{x_k=0} (1-f_k) \quad (k = 0,\ldots,i-1)$$

for every $x \in X_i$. Grouping now x's into pairs such that x_1^e and x_1^o are subsequent numbers $x_1^e < x_1^o$, x_1^e is even, $\mu(x_1^e)$ will differ from $\mu(x_1^o)$ only in one fit: f_0 and $1-f_0$, so

$$s_{0,1} = \mu(x_1^e) + \mu(x_1^o) = \prod_{x_k=1} f_k \cdot \prod_{x_k=0} (1-f_k) \quad (k = 1,\ldots,i-1)$$

Now making pairs of $s_{o,1,m}^e$ and $s_{o,1,m}^o$ so that they are subsequent and in x_1 bit b_1 is 0, we obtain

$$s_{1,m} = s_{o,1,m}^e + s_{o,1,m}^o = \prod_{x_k=1} f_k \cdot \prod_{x_k=0} (1-f_k) \quad (k = 2,\ldots,i-1)$$

etc. until

$$s_{i-1} = \sum_{X_i} \mu(x) = 1$$

is obtained as a product of zero factors. Q.e.d.

It is also clear that

$$\sum_{X_i} \mu(x) > 1$$

if $\wedge$ represents the min operator.

3. FURRY ARITHMETICS

3.1. Furry half addition

We face a very difficult problem if we intend to introduce binary operations especially addition among such numbers. The truth table of half addition is as follows:

TABLE 3. Truth table of half addition

A B	$S_H C_H$
0 0	0 0
0 1	1 0
1 0	1 0
1 1	0 1

By Table 3 we obtain similar marginal conditions for **extended half addition** as we had for the F^3 as it must be fulfilled for every fit position. Another restriction, analogous to P5 is proposed in addition:

$$A = 0.5,\ B = arbitrary \Rightarrow S_H = 0.5 \qquad (14)$$

(14) expresses the most ambiguous situation where the value of A is completely undefined. This also means that the result must be also ambiguous (with entropy 1). Here the "average value" of S and C must be taken as the resulting bits are also not informative.

In the above case a similar method as in Section 1.2 can be applied and the algebraic expressions of the sum and carry can be determined:

$$S_H = \alpha_0 + \alpha_1 A + \alpha_2 B + \alpha_3 AB$$
$$0 = \alpha_0 + \alpha_1.0 + \alpha_2.0 + \alpha_3.0 \Rightarrow \alpha_0 = 0$$
$$1 = \alpha_0 + \alpha_1.1 + \alpha_2.0 + \alpha_3.0 \Rightarrow \alpha_1 = 1$$
$$1 = \alpha_0 + \alpha_1.0 + \alpha_2.1 + \alpha_3.0 \Rightarrow \alpha_2 = 1$$
$$0 = \alpha_0 + \alpha_1.1 + \alpha_2.1 + \alpha_3.1 \Rightarrow \alpha_3 = -1$$

So S_H can be uniquely defined:

$$S_H = A + B - 2AB \qquad (15)$$

(15) also satisfies (14) as

$$0.5 + B - 2*0.5B = 0.5$$

Similarly

$$C_H = \beta_0 + \beta_1 A + \beta_2 B + \beta_3 AB$$
$$0 = \beta_0 + \beta_1.0 + \beta_2.0 + \beta_3.0 \Rightarrow \beta_0 = 0$$
$$0 = \beta_0 + \beta_1.1 + \beta_2.0 + \beta_3.0 \Rightarrow \beta_1 = 0$$
$$0 = \beta_0 + \beta_1.0 + \beta_2.1 + \beta_3.0 \Rightarrow \beta_2 = 0$$
$$1 = \beta_0 + \beta_1.1 + \beta_2.1 + \beta_3.1 \Rightarrow \beta_3 = 1$$

Hence we obtain

$$C_H = AB \qquad (16)$$

The above equations can be also generalized with arbitrary norms:

$$S_H = d(A \text{ s } B, A \text{ t } B) \ , \ C_H = A \text{ t } B \qquad (17)$$

We illustrate (15) and (16) by a numerical example:

$$\begin{array}{rrr} & 0.3 & 0.8 \\ + & 0.4 & 0.6 \\ \hline 0.3408 & 0.4984 & 0.4400 \end{array}$$

The half addition defined in (15) and (16) has some properties which resemble really on binary half addition. We list some of them in the next Statement.

Statement 2

Half addition defined by

$$S_H = d(A \text{ s } B, A \text{ t } B) \ , \ C_H = A \text{ t } B$$

is commutative and with algebraic connectives also associative.

Proof

The norms are commutative, so

$A \text{ s } B = B \text{ s } A$ and $A \text{ t } B = B \text{ t } A$

Hence

$$C_H(A,B) = A \text{ t } B = B \text{ t } A = C_H(B,A)$$

$d(X,Y)$ equals $X - Y$ if $X \geq Y$ and equals $Y - X$ if $X < Y$.

$$A \text{ s } B \geq A \text{ t } B$$

always holds, so

$$S_H(A,B) = d(A \text{ s } B, A \text{ t } B) = (A \text{ s } B) - (A \text{ t } B) =$$
$$= (B \text{ s } A) - (B \text{ t } A) = d(B \text{ s } A, B \text{ t } A) = S_H(B,A)$$

Commutativity is proven.

$$S_H((A,B),C) = (A + B - 2AB) + C - 2C(A + B - 2AB) =$$
$$= A + B + C - 2AB - 2AC - 2BC + 4ABC =$$
$$= A + (B + C - 2BC) - 2A(B + C - 2BC) = S_H(A,(B,C))$$
$$C_H((A,B),C) = ((AB)C) = (A(BC)) = C_H(A,(B,C))$$

Associativity for algebraic connectives is proven.

3.2. Furry addition

On similar basis also the **extended full addition** table and fuzzy full addition functions can be determined. Truth table of the full addition is to be found in Table 4.

To this table the according postulate of ambiguity (cf. (15)) is added, stating that any sum where one of the operands equals *0.5* will equal *0.5* as well (see Equ. 18).

TABLE 4. Truth table of full addition

A B C	S C
0 0 0	0 0
0 0 1	1 0
0 1 0	1 0
0 1 1	0 1
1 0 0	1 0
1 0 1	0 1
1 1 0	0 1
1 1 1	1 1

$$A = 0.5,\ B,C = arbitrary \Rightarrow S = 0.5, \qquad (18)$$

The linear equation system constructed on the basis of Table 4 and determining the sum function is as follows:

$$[\alpha_i]^T \mathbf{V}((A,B,C) = \mathbf{v}_j^T) = \mathbf{s}_j \ \ j = 0...8,$$

where

$$[\alpha_i] = \begin{bmatrix} \alpha_0 \alpha_0 & \dots \alpha_0 \\ \vdots & \vdots \\ \alpha_7 \alpha_7 & \dots \alpha_7 \end{bmatrix}$$

represents the coefficients belonging to the terms in

$$\mathbf{V} = [1, A, B, AB, C, AC, BC, ABC]^T$$

and

$$[\mathbf{v}_j] = \begin{bmatrix} 010 & & 01 \\ 001 & \dots & 11 \\ 000 & & 11 \end{bmatrix}, \ \mathbf{s}_j^T = [0\ 1\ 1\ 0\ 1\ 0\ 0\ 1],$$

obtained from column *S* of the Table.

Similarly the equation system defining the carry function is

$$[\beta_i]^T \mathbf{V}((A,B,C) = \mathbf{v}_j^T) = \mathbf{c}_j \; j = 0...8,$$

where

$$[\beta_i] = \begin{bmatrix} \beta_0 & \beta_0 & \cdots & \beta_0 \\ \vdots & & & \vdots \\ \beta_7 & \beta_7 & \cdots & \beta_7 \end{bmatrix}, \; \mathbf{c}_j^T = [0\;0\;0\;1\;0\;1\;1\;1],$$

obtained from column C of the Table.

Solution of the two linear equation systems results in the following:

$$\begin{aligned} S &= A + B + C - 2AB - 2AC - 2BC + 4ABC \\ C &= AB + BC + AC - 2ABC \end{aligned} \tag{19}$$

It is important to remark that (19) preserves the ambiguity principle (18) as

$$0.5 + B + C - 2*0.5B - 2*0.5C - 2BC + 4*0.5BC = 0.5$$

So (19) is the fundamental equation pair defining **furry addition** fit by fit.

We mention that the generalized representation using arbitrary norms is also possible but it is rather complicated. It is important that furry addition preserves some typical properties of crisp addition.

Statement 3

The result of furry addition with algebraic connectives does not depend on the sequence of terms.

Proof

It is obvious that both expressions are symmetrical in all the three operandi.

3.3. Furry subtraction?

An open problem is **furry subtraction** which can be defined by addition of the two's complement of the subtrahend. The first problem is the definition of **furry two's complement**.

One of the ways of calculating the two's complement is by generating first the one's complement or negation and then adding 1. So So the two's complement $\bar{A}$ of the furry number

$$A = \overline{a_{n-1}a_{n-2}\ldots a_1 a_0}$$

is the following:

$$\bar{A} = \overline{\bar{a}_{n-1}\bar{a}_{n-2}\ldots\bar{a}_1\bar{a}_0} + \overline{00\ldots01}, \tag{20}$$

where $\dotplus$ stands for furry addition and $^{-}$ denotes normal fuzzy negation $d(1,.)$.

Let us illustrate definition (20) by a numerical example:

$$A = \overline{0.4\ 0.6} \Rightarrow A^c = \overline{0.6\ 0.4}\ ,$$
$$A^c + 001_2 = \overline{0.76\ 0.52\ 0.6} = \bar{A}$$

Which are the properties of furry subtraction? Maybe the most important property of "crisp" subtraction is $A - A = 0$. Let us calculate now the sum of A and its two's complement $\bar{A}$:

$$\overline{0\ 0.4\ 0.6} + \overline{0.76\ 0.52\ 0.6} = \overline{0.2655\ 0.8439\ 0.5011\ 0.48}$$

Not considering the digits d_3 and d_2, the number represented by d_1 and d_0 is different from 0, it is a number near to absolute uncertainty i.e. $\overline{0.5\ 0.5}$. In general, the least significant fit of $X \dotplus \bar{X}$ will be

$$\Xi_0 = 2X_0 - 2X_0^2,$$

which expression assumes 0 if and only if $X_0 = 0$ or 1. Ξ_0 is near to 0.5 (i.e. absolute ambiguity or maximal entropy) if X_0 is also near to 0.5, according to corresponding postulate.

More generally, the following can be said:

Statement 4

When adding real fuzzy (not crisp) fits the ambiguity or entropy of the result is monotonously increasing.

Proof

For proving this statement let us take two real fuzzy fits $x = 0.5 + \varepsilon$, $y = 0.5 + \delta$, where

$$\varepsilon, \delta \in (0, 0.5),$$

i.e. both are really fuzzy. When adding them we obtain

$$s = x \dotplus y = (0.5 + \varepsilon) + (0.5 + \delta) - 2(0.5 + \varepsilon)(0.5 + \delta) =$$
$$= 0.5 + 2\varepsilon\delta,$$

where

$$2\varepsilon\delta < \min\{\varepsilon, \delta\}$$

and $\dotplus$ stands for furry half addition.

In general, supposing n addenda so that the fits are

$$f_i = 0.5 + \varepsilon_i,$$

further on that in all the addenda

$$|\varepsilon_i| > b > 0$$

the furry sum of f_i is

$$S_n = \sum_{i=1}^{n}{}^{f} (0.5+\varepsilon_i) \quad 0.5$$

at least with geometrical speed. (Σ^f stands for furry sum.)

Of course the above statement is applicable for the case where any number and its two's complement is added - except when crisp numbers are taken. So the series

$$q_i = q_{i-1} \dotplus \bar{q}_{i-1}, \; q_0 \neq 0,1$$

converges to 0.5 with geometrical speed.

Our conclusion is that furry subtraction (and also addition) defined in the above way leads to a fast loss of information in the result. It is an open question if by some other definition another type of extension of the binary arithmetical operations can be achieved.

3.4. Furry counting

It is interesting to check the extension of **counting**. Counting is nothing else but addition of 1. Using now (15) and (16) we have

$$s_0 = f_0 \dotplus 1 = f_0 + 1 - 2f_0 = 1 - f_0$$
$$c_0 = f_0.1 = f_0$$

So the entropy of the lowest significance bit never changes with counting: it wil alternate between f_0 $1 - f_0$. Of course, for s_1 we have

$$s_1 = f_1 + f_0 - 2f_1f_0$$
$$c_1 = f_1f_0$$

for which the above statement is valid.

The examination of furry numbers and furry arithmetics will be continued.

REFERENCES

[1] T. Yamakawa, T. Miki and F. Ueno: The Design and Fabrication of the Current Mode Fuzzy Logic Semi-Custom IC in the Standard CMOS IC

Technology. *Proc. 1985 ISMVL* IEEE, 1985.pp. 76-82

[2] T. Yamakawa: High-Speed Fuzzy Controller Hardware System. *Proc. of the 2nd Fuzzy System Symposium of the IFSA Japan Chapter.* Tokyo, 1986. pp. 122-130.

[3] T. Yamakawa:. A Simple Fuzzy Computer System Applying min & max Operations - A Challenge to 6th Generation Computer. *Proceedings of the 2nd IFSA World Congress, Vol. 2.,* Tokyo, 1987. pp. 827-830.

[4] P.P. Bonissone: Summarizing and Propagating Uncertain Information with Triangular Norms. *Int. J. of Approximate Reasoning 1987. 1.* pp. 71-101.

[5] K. Hirota - K. Ozawa: Fuzzy Flip-Flop as a Basis of Fuzzy Memory Modules. In: *M.M. Gupta - T. Yamakawa (eds.): Fuzzy Computing: Theory, Hardware, Realization and Applications.* North Holland, Amsterdam - New York, 1988.

[6] K. Hirota - K. Ozawa: Fuzzification of Flip-Flop Based on Various Logical Operations. *Bulletin of the College of Eng., Hosei Univ., Koganei-city. 23 (March 1987).* pp. 69-94.

[7] K. Hirota - K. Ozawa: Concept of Fuzzy Flip Flop. *Proceedings of the 2nd IFSA World Congress, Vol. 2.,* pp. 556-559.[8] L.T. Kóczy: Vectorial I-fuzzy Sets. In: *M.M. Gupta - E. Sanchez (eds.): Approximate Reasoning in Decision Analysis.* North Holland, Amsterdam - New York, 1982. pp. 151-156.

[9] L.T. Kóczy - C. Magyar: On the Minimal Axiomatic System of I-fuzzy Structure. *BUSEFAL Automne 1987.* pp. 19-31.

[10] B. Kosko: Fuzzy Entropy and Conditioning. *Information Sciences vol. 40 (1986).* pp. 165-174.

[11] L. T. Kóczy - K. Hirota - K. Ozawa: Knowledge representation and accumulation by fuzzy flip-flops. In: *M.M. Gupta - Z.A. Azmi (eds.):* Contributed volume to appear in 1989.

[12] K. Hirota, L.T. Kóczy and K. Ozawa: Fundamental logic in fuzzy flip-flops. In: *Proc. Nineteenth Annual Pittsburgh Conference on Modeling and Simulation.* University of Pittsburgh School of Engineering, Pittsburgh, 1988. pp. 2165-2168.

[13] K. Hirota, L.T. Kóczy and K. Ozawa: Discrete Mode Algebraic Fuzzy Flip-Flop Circuit. *Proc. of the Int. Workshop on Fuzzy System Applications.* Kyushu Institute of Technology, Iizuka, 1988. pp. 39-40.

[14] L. T. Kóczy: Some remarks concerning fuzzy digital circuits. *H. R. Hansen - W. H. Janko (eds.): 2nd Joint IFSA-EC EURO-WG Workshop on "Progress in Fuzzy Sets in Europe" Abstracts.* University of Economical Science, Vienna, 1988. pp. 61-65.

APPROXIMATE REASONING IN THE POSSIBILITY THEORY FRAMEWORK : where are we practically ?

Roger MARTIN-CLOUAIRE
Station de Biométrie et Intelligence Artificielle
INRA, B.P. 27
31326 CASTANET-TOLOSAN CEDEX
FRANCE

ABSTRACT : This paper discusses contemporary issues associated with the use of approximate reasoning techniques in knowledge-based systems relying on the theory of possibility. It aims at :

- providing lines along which comparison of existing systems can be done ;
- pointing out directions toward which future systems should evolve.

1. INTRODUCTION

Representational and inferential capabilities based on imprecise , uncertain, incomplete or inconsistent information are becoming more important in the design, implementation and operation of knowledge-based systems. Since the advent of the first expert systems several approaches for making inference from available information have been developed. This general problem has benefited from important contribution coming out of research conducted in the possibility theory framework. Unfortunately, the results from this contribution have not yet received much attention outside the fuzzy set community and may not have spread enough within this community itself. One of the reason may be that the great variety and complexity of the facets constituting this problem are confusing people. Therefore, clarifying the picture is a need and is modestly attempted here (see also Chapter 7 of [Zim87] for a survey).

This paper discusses contemporary issues associated with the use of approximate reasoning techniques in rule-based systems relying on the theory of possibility. We first examine the best known of these techniques with respect to their capabilities in processing

W. H. Janko et al. (eds.), Progress in Fuzzy Sets and Systems, 115–124.

"imperfect knowledge". Then we provide a perspective concerning some general approximate reasoning problems whose solutions have not yet received a satisfactory implementation (if any). Thus, the paper is pursuing a twofold goal :

- providing lines along which comparison of existing systems can be done ;
- pointing out directions toward which future systems should evolve.

2. EXISTING SYSTEMS

In this section, we propose a little review of some existing systems along a set of discriminating features. No claim of exhaustivity for either the systems or the features is made.

Differentiation between systems can be done by examining what kind of knowledge they are able to represent and process effectively. Indeed, beside the external aspect of rules and facts that constitute the knowledge base, one must observe how these systems behave in the typical four step process characterizing an inference because it is difficult to understand what is really represented without taking into account how information is used. The four steps under consideration are :

i) evaluation of the individual conditions of a rule with respect to corresponding facts ;
ii) aggregation of the above elementary evaluations ;
iii) deduction by the combination of step ii) with conditional information representing the antecedent-consequent relationship ;
iv) fusion of the result of step iii) with other related items of information.

It is convenient, for the present discussion, to consider deductive reasoning under the form of the basic pattern that follows :

$$\begin{array}{l} \text{if } p \text{ then } q \\ \quad p' \\ \hline \qquad q' \end{array} \qquad (1)$$

where p, q, p' and q' are propositions, p' is supposed to match p somehow and q' is what can be deduced from the pieces of information above the line. The sophistication of systems clearly depends on whether or not they support inferences where p, q and p' can be vague propositions. Let us recall that a proposition is uncertain when one cannot definitely state it is true of false. A proposition is imprecise if the single-valued parameter it aims to describe is incompletely specified in the sense that more than one value is possible. Vagueness is a particular kind of imprecision where the possible values of the parameter lack clear or crisp boundaries. Thus, uncertainty refers to the truth of the proposition whereas imprecision pertains to the content of the proposition or, in other words, its meaning.

2.1 Treating uncertainty

Some of the existing systems (e.g. RUM [Bon87], FESS [Hal86]) are addressing the issue of approximate reasoning as a problem of management of uncertainty only (where the proposition involved in pattern (1) are non-vague) because the applications they aim to do not require meaning computation. For more sophisticated systems, uncertainty management techniques are provided explicitly of implicitly as special case of more powerful capabilities.

The uncertainty about the truth of a proposition is often represented by one (e.g. [Hal87] [Buc86]) or two numbers (e.g. [Mar85], [Bon87]) belonging to [0,1]. In the case two numbers are used (e.g. necessity and possibility degrees in SPII [Mar86]) one can express to what extent it is certain that p is true and to what extent the contrary of p (denoted ¬p) is true. It makes it possible to represent ignorance (which is a lack of support in favor of p as well as ¬p) and to distinguish it from total uncertainty (where the support in favor of p equals the support in favor of ¬p and is right in between total absence of support and full support). Using a single number makes this distinction impossible except if uncertainty has a probabilistic nature since in such a case support in favor of p determine completely the support in favor of ¬p. Some authors have interpreted such a single number as an intermediate truth value. But this approach is strongly criticizable because it does not make sense to say that a precise proposition is more or less true. It can only be either true or false although one may not have full confidence in either its truth or its falsity.

The representation of the uncertainty on the inferential relationship between p and q in pattern (1) is another characteristic along which systems differ one from the other. Again, either one or two numbers may be used. The case of two numbers (e.g. [Leb86], [Bon87]) aims at expressing to what extent it is necessary (compulsory) and sufficient to have p true in order to deduce that q is true as well. The necessity aspect answers the purpose of deducing to what extent q is false when there is support in favor of ¬p. On the other hand, the sufficiency aspect serves in evaluating to what extent q is true when there is support in favor of p. Systems using a single degree (e.g. [Muk87], [Uma87]) usually deal with the sufficiency aspect only.

Representation issues concerning uncertainties on facts and rules have immediate consequences on the steps i) and iii) of the inference process. Steps ii) and iv) are basically two kinds of aggregation operations. Several existing systems (e.g. [Bon87], [Buc86]) provides many different possibilities (using families of t-norm and t-conorm operators) for performing these combinations of uncertain information. Some (e.g. [Leb87]) have so far adopted the safer position that consists in using 'min 'and 'max' operators only, thus keeping the meaning given to the the degrees of uncertainty globally in agreement with the underlying mathematical framework i.e. possibility theory.

2.2 Uncertain conclusion together with vagueness in conditions and facts

We consider here situations where p and p' in (1) are of the form X is A and X is A' respectively where A and A' are fuzzy sets. The step i) of the inference process has to be extended in order to accommodate the imprecision of information conveyed by p and p'. Basically, two approaches are encountered. The first one [Uma87] involves the computation of the compatibility of A and A' i.e. CP(A ; A'). This compatibility is a fuzzy degree of truth (intermediate truth makes sense in this case) of X is A given that X is A'. The processing of such fuzzy degrees of truth (i.e fuzzy sets in [0,1]) in the four steps of inference may be very costly from a computational point of view. For this reason, their use is done in a highly constrained manner. For instance, Umano [Uma87] deals with fuzzy truth values represented by discrete possibility distributions involving few points. As another example, in MILORD [God87] only positive truth values are used (i.e. falsity is not represented). Moreover, any truth value is assumed to belong to a set of nine values (via an approximation if needed). All logical operations involved in the four steps of inference can then be expressed by "truth tables" on the nine truth value set. Although, in MILORD the p and p' of Pattern (1) may be fuzzy no meaning computation is really needed because p is expressed under the form "X is M is ∂" and p' by "X is M is ß" where ∂ and ß are fuzzy truth values. Thus, the only difference is on the truth value qualifying "X is M" so that computation can be carried on at the level of uncertainty only. If p where allowed to be of the form "X is N is ß" with N different from M then meaning processing (involving explicit consideration of the fuzzy sets M and N) would be necessary.

The second approach (see for instance [Mar85], [Sou86]) uses necessity and possibility degrees (i.e. N(A ; A') and Π(A ; A')) for expressing to what extent p' matches p. It is important to keep in mind that these two degrees convey information also contained in the fuzzy compatibility degrees considered in the other approach. Indeed, one can always compute N(A ; A') and Π(A ; A') from CP(A ; A'). Thus, the two approaches clearly share the same mold but the second one is not confronted to any inefficiency problem. When the similarity between the conditions and facts is expressed via necessity and possibility degrees the inference process can be carried on in the same way than with rules involving precise conditions only. Thus, this approach nicely copes with rules having both precise and vague conditions.

2.3 General framework for managing imprecision and uncertainty

Zadeh's general modus ponens corresponds to the pattern (1) where p , q, p' and q' are of the form "X is A", "Y is B","X is A'" and "Y is B'" respectively. Approximate reasoning in this scheme is seen as a problem of meaning computation or constraint propagation. Indeed the meaning of q' (i.e the restriction on the possible values of Y) is obtained by combining (via a non-linear programming technique) the restriction on the possible values of X with the semantic content of the rule expressing the X-Y relationship. This powerful technique is used for instance in SPII [Leb87] and CARDIAG-2 [Adl86]. In the latter case, the involved fuzzy sets are on discrete universes and the X-Y relationship is explicitly defined by the user rather than induced from a loose specification conveyed by the rule.

Usually, the fuzzy relation expressing the X-Y relationship is obtained by means of a multiple-valued implication connective → such that $\mu_{X\text{-}Y}(s,t) = \mu_A(s) \rightarrow \mu_B(t)$.
The generalized modus ponens permits to compute B' as in formula (2)where * is a conjunction operator.

$$\mu_{B'}(t) = \sup_s \mu_{A'}(s) * \mu_{X\text{-}Y}(s,t) \qquad (2)$$

The different possible choices for * and -> lead to many variants (see [Mar88] for some examples) but, if we want B' to be equal to B when A' is equal or included in A then the choice of → cannot be done independently of the one concerning * [Dub84]. These choices of operators correspond to different interpretations of the rule " if X is A then Y is B". Other interpretations are possible if we consider the rule under it contraposive form (i.e. "if Y is not B then X is not A") since, in fuzzy logic, a rule is not necessarily equivalent to its contrapositive form. For instance, if the rule " if X is A then Y is B" is used in its contrapositive form together with the 'min' conjunction operator and the Gödel implication function then the conditional piece of knowledge actually represented tells something like "the more X is A the more certain the proposition Y is B" [Bui86].

The X-Y dependency is often described via a collection of rules rather than a single one. More specific results are obtained if the rules are combined before being used in inference [Mar88]. So far, no known system performs this prior combination.

The main reason why the generalized modus ponens has not been used more extensively in practical systems is that its straightforward implementation gives very poor performance. However, a satisfactory use of this technique can be reached with continuous parametrized possibility distributions and some approximation procedures. The technique developed in [Mar88] permits to see the generalized modus ponens as a deductive device running in a manner similar to the four step inference earlier mentioned in this paper although it does not if considered according to its basic definition.

2.4 Additional discriminating features

In the step ii) of inference with rules having precise conclusions one has to perform an aggregation of the levels of satisfaction of the individual conditions by the corresponding facts. A useful improvement is obtained if one take into account the notion of importance of conditions (e.g. that a condition is more important than another one or that a condition need not be fully satisfied in order to apply the rule). A technique for incorporating importance in conjunction and disjunction operations performed with necessity and possibility degrees has been developed by Dubois et al.[Dub88] and implemented in late versions of SPII [Mar86]. It is not clear in the literature whether other systems have a similar feature.

Conditions of rule may contain numerical predicates. Systems like FLOPS [Buc86] or SPII can evaluate to what extent a given fuzzy number is greater than another one.

Some inference engines allow numerical computation in the antecedent or consequent part of rules. Few fuzzy inference engines can do the same with fuzzy numbers. SPII [Leb86] supports computation in conclusions provided the involved variables are non-interactive.

Fusion of information coming from different sources is another problem that has received different solutions across existing systems. So far no implementation seems sufficiently satisfactory and some use very ad-hoc combination operations for procedural purpose only [Adl86].

Related to the problem of combination is the one of controlling the search of the system toward a goal. In order to avoid exhaustive search some have introduced thresholding techniques that give quantitative flavor to degrees of uncertainty. Rules may be invoked according to the amount of their uncertainty or with respect to their specificity or any other criteria referring to imprecision and uncertainty.

Finally, another line of comparison of existing systems concerns their abilities in explaining results of uncertain reasoning. This often involves linguistic approximation techniques in order to ensure proper communication in linguistic terms.

3. GOING FARTHER

3.1 Representing more information

Future improvement of fuzzy inference engines should come from progresses in knowledge representation. It has been pointed out that the generic rule "if X is A then Y is B" can be interpreted in different ways. However, it is important to note that the generalized modus ponens, as has been considered so far in existing systems, is always oriented toward one particular understanding of the rule (we do not consider the contrapositive form here). Basically, it tells that the derived B´ restricts the values in the domain of Y which are in relation (via the rule) with at least one value in A´. As shown in [Dub87], there is a dual interpretation which permits to obtain the constraint on the values that are in relation (again via the rule) with all values of the domain of X more or less compatible with A´. Thus, the second acceptation of the rule gives a more restrictive possibility distribution (that is contained in the one resulting from the first one). Both interpretations may be useful in practice.

In addition to the above new rules one could also need, for instance, rules of the kind "the more X is A, the more Y is B" [Pra87]. Such a kind of rule describes in a qualitative manner the variation or the variable Y in terms of the variation of X.

So far, fuzzy sets used in most systems have a disjunctive meaning (i.e. they restrict the more or less possible values of a variable taking a single value). The conjunctive interpretation (i.e. where the described variable may have several values) should be taken into account too.

3.2 Better combination of imprecise and uncertain information

Given several items of information concerning a particular variable and coming from different sources one is confronted with the problem of synthesizing the pieces into a single item. Most of the time the combination is computed by performing a conjunction. However the combination process must depend on what has to be combined. In some cases, a disjunction or a compromised combination may be the good choice. In other cases the combination process must discard some items [Dub87b].

Future systems should handle the combination problem with more attention. This may have deep consequences on system architecture because solving the problem of combination requires specific reasoning activities.

3.3 Merging different techniques

Knowledge-based systems often have to face problems of default reasoning. The techniques embedded in current systems make it possible to solve some of these problems but others are better handled by purely symbolic approaches or syllogistic reasoning methods [Zad87]. Future systems must provide a way for merging the different techniques needed for non-monotonic reasoning.

4. REFERENCES

Adl86 Adlassnig, Scheithauer, W., Kolarz, G. Fuzzy medical diagnosis in a hospital. In Fuzzy Logic in Knowledge Engineering (H. Prade, C. Negoita, eds.), Verlag TÜV Rheinland, Köln, 275-294, 1986.

Bon87 Bonissone, P., Gans, S., Decker, K. RUM : a layered architecture for reasoning with uncertainty. Proc. IJCAI-87, Milan, 891-898, 1987.

Buc86 Buckley, J., Siler, W., Tucker, D. FLOPS, a fuzzy expert system : applications and perspectives. In Fuzzy Logic in Knowledge Engineering (H. Prade, C. Negoita, eds.), Verlag TÜV Rheinland, Köln, 256-274, 1986.

Bui86 Buisson, J-C., Farreny, H., Prade, H. Dealing with imprecision and uncertainty in the expert system DIABETO-III. Proc. 2nd Int. Colloq. on Artificial Intelligence, Marseille, Dec.1-5, Hermès, Paris, 705-721,1986.

Dub84 Dubois, D., Prade, H. Fuzzy logics and the generalized modus ponens revisited. Cybernetics & Systems, 15, n°3-4, 293-331, 1984.

Dub85 Dubois, D., Prade, H. (with the collaboration of H. Farreny, R. Martin-Clouaire, C. Testemale) Théorie des Possibilités. Application à la Représentation des Connaissances en Informatique. Masson, Paris, 1985. English updated version : Possibility Theory - an Approach to Computerized Processing of Uncertainty, Plenum Press, New York, 1988.

Dub87 Dubois, D., Prade, H. Upper and lower images of a fuzzy set induced by a fuzzy relation. To appear in Fuzzy Expert Systems, (A. Kandel, ed.), Addison-Wesley, 1987.

Dub87b Dubois, D., Prade, H. On the combination or uncertain or imprecise pieces of information in rule-based systems - a discussion in the framework of possibility theory. To appear in Int. J. of Approximate Reasoning, 1988.

Dub88 Dubois, D., Prade, H., Testemale, C. Weighted fuzzy pattern matching. To appear in Fuzzy Sets & Systems, 1988.

God87 Godo, L., Lopez de Mantaras, R., Sierra, C., Verdaguer A. Managing linguistically expressed uncertainty in MILORD application to Medical diagnosis. Proc. Int. Workshop on Expert Systems and their Applications, Avignon, 571-596, 1987.

Hal86 Hall, L., Kandel, A. Designing Fuzzy Expert Systems. Verlag TÜV Rheinland, Köln,1986. Reviewed by R. Martin-Clouaire in Fuzzy Sets & Systems, 23, 408-411, 1987.

Leb87 Lebailly, J., Martin-Clouaire, R., Prade, H. Use of fuzzy logic in a rule-based system in petroleum geology. In Approximate Reasoning in Intelligent Systems, Decisions and Control, (E. Sanchez, L.A. Zadeh, eds.), Pergamon Press,125-144, 1987.

Mar85 Martin-Clouaire, R., Prade, H. On the problems of representation and propagation of uncertainty in expert systems. Int. J. Man-Machine Studies, 22, 251-264, 1985.

Mar86 Martin-Clouaire, R., Prade, H. SPII-1, a simple inference engine capable of accommodating both imprecision and uncertainty. In Computer-Assisted Decision Making, (G. Mitra, ed.), North Holland, 117-131, 1986.

Mar88 Martin-Clouaire, R. Semantics and computation of the generalized modus ponens: the long paper. To appear in Int. J. of Approximate Reasoning, 1989.

Muk87 Mukaidono, M., Shen, Z., Ding, L. Fuzzy PROLOG. Proc. IFSA-87, Tokyo, 452-455, 1987.

Pra87 Prade, H. Raisonner avec des règles d'inférence graduelle - une approche basée sur les ensembles flous. Submitted, 1987.

Sou86 Soula, G., Vialettes, B., San Marco, J-L., Thirion, X., Roux, M. PROTIS : a fuzzy expert system with medical applications. In Fuzzy Logic in Knowledge Engineering (H. Prade, C. Negoita, eds.), Verlag TÜV Rheinland, Köln, 295-310, 1986.

Uma87 Umano, M. Fuzzy set PROLOG. Proc. IFSA-87, Tokyo, 750-753, 1987.

Zad87 Zadeh, L.A. A computational theory of dispositions. Int. J. of Intelligent Systems, 2, 39-63, 1987.

Zim87 Zimmermann, H.J. Fuzzy Sets, Decision Making and Expert Systems , Kluwer Academic Publishers, 1987.

ON CALCULATING THE COVARIANCE IN THE PRESENCE OF VAGUE DATA

Klaus Dieter Meyer
AEG Aktiengesellschaft
– Forschungsinstitut Berlin –
Holländerstr. 31–34
D-1000 Berlin 51
Federal Republic of Germany

Rudolf Kruse
Technische Universität Braunschweig
Institut für Betriebssysteme und Rechnerverbund
Bültenweg 74–75
D-3300 Braunschweig
Federal Republic of Germany

ABSTRACT. In this paper we consider the problem of covariance-analysis in the presence of vague data. We have to do with two random mechanisms and are interested in the covariance of them. We obtain not precise but only vague observations, therefore we can only expect a vague value for the covariance. The vague observations can be described by pairs of fuzzy sets of the real line being considered as realizations of a pair of so called fuzzy random variables (f.r.v.'s). We show how to calculate the covariance of two finite f.r.v.'s. Being a fuzzy set, the covariance is the fuzzy perception of the covariance of the two underlaying random mechanisms. In practice we do not know the common probabilty distribution of the two f.r.v.'s which are needed for calculating the fuzzy covariance. We show that an estimator for this distribution yields a strong consistent estimator for the covariance with respect to a suitable metric. As a special case of a pair of f.r.v.'s, the covariance of two random sets can be calculated and estimated, resp.

1. Fuzzy sets and some properties

Fuzzy sets of the real line turned out to be an appropriate tool to represent vague data by generalizing other approaches (compare Kruse & Meyer [4]). Following Zadeh [8] we define:

Definition 1.1:
A *fuzzy set* of the real line is characterized of its membership function

$$\mu : I\!R \to [0,1].$$

W. H. Janko et al. (eds.), Progress in Fuzzy Sets and Systems, 125–133.

An important tool for dealing with fuzzy sets are their α-level sets and their strong α-cuts.

Definition 1.2:
For a fuzzy set μ and for $\alpha \in [0,1]$ we define

$$\mu_{\overline{\alpha}} \stackrel{d}{=} \{x \in I\!R \mid \mu(x) \geq \alpha\} \quad (\alpha\text{-level-set})$$

$$\mu_{\alpha} \stackrel{d}{=} \{x \in I\!R \mid \mu(x) > \alpha\} \quad (\text{strong } \alpha\text{-cut})$$

As fuzzy sets represent vagueness, we can restrict ourselves to fuzzy sets with a rather simple membership function. We only consider fuzzy sets of the following class:

Definition 1.3:
Let $Q(I\!R)$ denote the class of fuzzy sets μ fulfilling the following conditions:

(i) there exists $x \in I\!R$ with $\mu(x) = 1$

(ii) $\inf \mu_0 > -\infty$ and $\sup \mu_0 < +\infty$

(iii) $\mu_{\overline{\alpha}}$ is closed for all $\alpha \in [0,1]$.

Myiakoshi & Shimbo [6] introduced — as a generalization of the tools α-level sets and strong α-cuts — the concept of a set representation of a fuzzy set.

Definition 1.4:
Let $\mu \in Q(I\!R)$. $\{A_\alpha \mid \alpha \in (0,1)\}$ is called a *set representation* of μ if and only if

(i) $0 < \alpha < \beta < 1$ implies $A_\beta \subseteq A_\alpha$

(ii) for all $x \in I\!R : \mu(x) = \sup \{\alpha \cdot I_{A_\alpha}(x) \mid \alpha \in (0,1)\}$ where I_{A_α} denotes the indicator function of A_α.

The strong α-cuts and the α-level sets are the 'smallest' and the 'largest' set representations. More precisely:

Theorem 1.5:
Let $\mu \in Q(I\!R)$. Then it holds: $\{A_\alpha \mid \alpha \in (0,1)\}$ is a set representation of μ if and only if $\mu_\alpha \subseteq A_\alpha \subseteq \mu_{\overline{\alpha}}$ holds for all $\alpha \in (0,1)$.

2. Fuzzy random variables

The concept of a *fuzzy random variable* (f.r.v.) is an approach for handling two different kinds of uncertainty — vagueness and randomness. Fuzzy random variables have been introduced by Kwakernaak [5]. For a comparism with other approaches we refer to Kruse & Meyer [4] and Puri & Ralescu [7].

Let $(\Omega, \mathcal{A}, P)$ be a probability space. We consider a random mechanism on this probability space. We take into account that the random mechanism is disturbed by influences of the environment or measure errors, e.g. This disturbances are described by the influence of annother probability space $(\Omega', \mathcal{A}', P')$. We assume that each possible error can occur and also different errors can occur simultaneously. Therefore we demand that $(\Omega', \mathcal{A}', P')$ is 'rich enough', more precisely: for all nonnegative reals a, b, and c there exists three sets $A \in \mathcal{A}'$, $B \in \mathcal{A}'$ and, $C \in \mathcal{A}'$ being pairwise disjoint such that $P'(A) = a$, $P'(B) = b$, and $P'(C) = c$.

Our random mechanism depends on the product probability space of these two spaces being denoted by $(\Omega \times \Omega', \mathcal{A} \otimes \mathcal{A}', P \otimes P')$. It can be described by a usual random variable on $(\Omega \times \Omega', \mathcal{A} \otimes \mathcal{A}', P \otimes P')$. Therefore we have to deal with randomness. We cannot observe the realizations of this random variable directly but can only obtain vague observations. So we have to do with vagueness in addition to randomness. The vague observations can be described by fuzzy sets of the real line and are interpreted as realizations of a fuzzy random variable (f.r.v.). A f.r.v. is a vague perception of a usual random variable, on the other hand the underlaying random variable is called the *original* of the f.r.v. which realizations we observe.

As we cannot observe the influence of the second probability space $(\Omega', \mathcal{A}', P')$, i.e. the influences of the environment or measure error, e.g., directly, the f.r.v. only depends on $(\Omega, \mathcal{A}, P)$. Therefore a f.r.v. is a mapping $X : \Omega \to Q(\mathbb{R})$. Its original is unknown, we only assume that its expected value exists. Therefore the set of all possible originals is $\widetilde{X}$, the set of all random variables on $(\Omega \times \Omega', \mathcal{A} \otimes \mathcal{A}', P \otimes P')$ with existing expected value.

Let $X : \Omega \to Q(\mathbb{R})$ be a f.r.v. and let $U : \Omega \times \Omega' \to \mathbb{R}$ be a possible original. By fuzzy logic, we can derive the acceptability for the statement 'U is original of X' and obtain:

$$\mu_X(U) = \inf\{X_\omega[U(\omega, \omega')] \mid \omega \in \Omega, \omega' \in \Omega'\}.$$

As we only obtain vague observations, it is not sensitive to distinguish between an infinite number of vague observations. Therefore we restrict ourselves to finite f.r.v.'s. We define analogously to classical statistics:

Definition 2.1:
A mapping $X : \Omega \to Q(\mathbb{R})$ is called a *finite* f.r.v. with the codomain $\{\mu_1, \ldots, \mu_n\}$ if and only if

(i) $\{\omega \in \Omega \mid X(\omega) = \mu_i\} \in \mathcal{A} \quad (i = 1, \ldots, n)$

(ii) $P(X = \mu_i) > 0 \quad (i = 1, \ldots, n)$

(iii) $\sum_{i=1}^{n} P(X = \mu_i) = 1$.

A finite random set is a special case of a finite f.r.v. A random set L is a mapping $L : \Omega \to \wp(\mathbb{R})$ ($\wp(\mathbb{R})$ denotes the power set of $\mathbb{R}$). If we identify a set $A \subseteq \mathbb{R}$ with its indicator function I_A, the values of a random set are fuzzy sets.

We can transfer our considerations to multi-dimensional random mechanisms and their fuzzy perceptions and can define the notion fuzzy random vector.

Definition 2.2:
A mapping $(X,Y) : \Omega \to [Q(\mathbb{R})]^2$ is called a *finite fuzzy random vector* with the codomain $\{\mu_1,\dots,\mu_m\} \times \{\nu_1,\dots,\nu_n\}$ if X and Y are two finite f.r.v. with the codomains $\{\mu_1,\dots,\mu_m\}$ and $\{\nu_1,\dots,\nu_n\}$, resp.

3. A set representation of the covariance

In the following, let $(X,Y) : \Omega \to [Q(\mathbb{R})]^2$ denote a finite fuzzy random vector. We want to define and to calculate the covariance of X and Y as the fuzzy perception of the underlaying two random mechanism.

The extension principle of Zadeh [9] has been developed with the help of fuzzy logic. It turned out to be a useful procedure for transfering numerical operations from usual numbers to fuzzy sets as well as transfering statistical notions from usual to fuzzy random variables. For calculating a parameter of a f.r.v. we consider all possible originals of a f.r.v. and calculate on the one hand that parameter of the original and on the other hand the acceptability of this random variable to be the original of the f.r.v.

Using this approach, the expected value, the variance and the distribution function of f.r.v.'s was introduced (compare Kruse [2] and Kruse and Meyer [4]).

Being applied to the covariance, the extension principle leads us to the following definition:

Definition 3.1:
The fuzzy set $\operatorname{cov}(X,Y)$ is defined by its membership function

$$\operatorname{cov}(X,Y)(t) \stackrel{\mathrm{d}}{=} \sup\left\{\min[\mu_X(U),\mu_Y(V)] \mid U \in \widetilde{X}, V \in \widetilde{X}\right\}, \quad t \in \mathbb{R}.$$

$\mu_X(U)$ is the acceptability that U is original of X and $\mu_Y(V)$ that of V being original of Y. $\min[\mu_X(U),\mu_Y(V)]$ is the acceptability that a pair (U,V) of usual random variables is original of the vector (X,Y).

In the following let $\{\mu_1,\dots,\mu_m\} \times \{\nu_1,\dots,\nu_n\}$ contain the codomain of (X,Y). We allow $P(X = \mu_i \text{ and } Y = \nu_j) = 0$ for some $(i,j) \in \{1,\dots,m\} \times \{1,\dots,n\}$.

Define $\Gamma_i = \{\omega \in \Omega \mid X_\omega = \mu_i\}$ for $i \in \{1,\dots,m\}$ and $\Sigma_i = \{\omega \in \Omega \mid Y_\omega = \nu_j\}$ for $j \in \{1,\dots,n\}$ $\Gamma_i \in \mathcal{A}$ and $\Sigma_j \in \mathcal{A}$ is valid.

In order to calculate $\operatorname{cov}(X,Y)$ we start searching a set representation of this fuzzy set. The level sets of the codomains of X and Y are an important help for this task. We know that these level sets are compact subsets of the real line. More precisely:

Theorem 3.2:
Define for $\alpha \in (0,1)$

$$X_\alpha \stackrel{\mathrm{d}}{=} \{U \in \widetilde{X} \mid U(\omega,\omega') \in (\mu_i)_{\overline{\alpha}} \text{ holds for all } i \in \{1,\ldots,m\},\ \omega \in \Gamma_i,\ \omega' \in \Omega'\}$$
$$Y_\alpha \stackrel{\mathrm{d}}{=} \{V \in \widetilde{X} \mid V(\omega,\omega') \in (\nu_j)_{\overline{\alpha}} \text{ holds for all } j \in \{1,\ldots,n\},\ \omega \in \Sigma_j,\ \omega' \in \Omega'\}$$
$$C_\alpha \stackrel{\mathrm{d}}{=} \{t \in I\!R \mid \exists\, U \in X_\alpha, \exists\, V \in Y_\alpha \text{ mit } \operatorname{cov}(U,V) = t\}.$$

Then $C_\alpha\{\alpha \in (0,1)\}$ is a set representation of $\operatorname{cov}(X,Y)$.

By Theorem 3.2, the problem of calculating $\operatorname{cov}(X,Y)$ is reduced to that of calculating the C_α's. When calculating $\inf C_\alpha$ and $\sup C_\alpha$ we find that these values only depend on the 'edges' of the level sets of the codomain of (X,Y). We formulate:

Theorem 3.3:
Define for $A \in \mathcal{A} \otimes \mathcal{A}'$, $B \in \mathcal{A} \otimes \mathcal{A}'$ the usual random variables $U_A \in \widetilde{X}$ and $V_B \in \widetilde{X}$ by

$$U_A(\omega,\omega') \stackrel{\mathrm{d}}{=} \begin{cases} \sup(\mu_i)_{\overline{\alpha}} & \text{if } (\omega,\omega') \in A \quad \text{and } \omega \in \Gamma_i \quad (i = 1,\ldots,m) \\ \inf(\mu_i)_{\overline{\alpha}} & \text{if } (\omega,\omega') \in \Omega \times \Omega' \setminus A \quad \text{and } \omega \in \Gamma_i \quad (i = 1,\ldots,m) \end{cases}$$
$$V_B(\omega,\omega') \stackrel{\mathrm{d}}{=} \begin{cases} \sup(\nu_j)_{\overline{\alpha}} & \text{if } (\omega,\omega') \in B \quad \text{and } \omega \in \Sigma_j \quad (j = 1,\ldots,n) \\ \inf(\nu_j)_{\overline{\alpha}} & \text{if } (\omega,\omega') \in \Omega \times \Omega' \setminus B \quad \text{and } \omega \in \Sigma_j \quad (j = 1,\ldots,n). \end{cases}$$

We omit the details of the further calculation of $\inf C_\alpha$ and $\sup C_\alpha$. Both calculations yield to a minimum and a maximum, resp., of an algebraic expression of the 'edges' of the level sets. After having done this calculation a further investigation shows that C_α is a compact interval. We summarize our results in the following theorem:

Theorem 3.4:
Define $\Pi_{m,n} \stackrel{\mathrm{d}}{=} \{(p_{11},\ldots,p_{mn}) \in [0,1]^{m\cdot n} \mid \sum_{i=1}^{m}\sum_{j=1}^{n} p_{ij} = 1\}$.
Define for $\alpha \in (0,1)$, $(p_{11},\ldots p_{mn}) \in \Pi_{m,n}$, $T_i \subseteq \{1,\ldots,m\} \times \{1,\ldots,n\}$ $(i = 1,2,3,4)$, $(i,j) \in \{1,\ldots,m\} \times \{1,\ldots,n\}$, $t \in [0,1]$:

$$\begin{aligned} & f_{\alpha,p_{11},\ldots,p_{mn}}[T_1,T_2,T_3,T_4,(i,j)](t) \stackrel{\mathrm{d}}{=} \\ & \quad [\inf(\mu_i)_{\overline{\alpha}}\sup(\nu_j)_{\overline{\alpha}} - \sup(\mu_i)_{\overline{\alpha}}\inf(\nu_j)_{\overline{\alpha}}] \cdot p_{ij} \cdot t \\ & \quad + \sum_{(k,l)\in T_1} \sup(\mu_k)_{\overline{\alpha}}\sup(\nu_l)_{\overline{\alpha}} p_{kl} + \sum_{(k,l)\in T_2} \sup(\mu_k)_{\overline{\alpha}}\inf(\nu_l)_{\overline{\alpha}} p_{kl} \\ & \quad + \sum_{(k,l)\in T_3} \inf(\mu_k)_{\overline{\alpha}}\sup(\nu_l)_{\overline{\alpha}} p_{kl} + \sum_{(k,l)\in T_4} \inf(\mu_k)_{\overline{\alpha}}\inf(\nu_l)_{\overline{\alpha}} p_{kl} \\ & \quad - \Big\{[\inf(\mu_i)_{\overline{\alpha}} - \sup(\mu_i)_{\overline{\alpha}}] \cdot p_{ij} \cdot t + \sum_{(k,l)\in T_1\cup T_2} \sup(\mu_k)_{\overline{\alpha}} \cdot p_{kl} + \sum_{(k,l)\in T_3\cup T_4} \inf(\mu_k)_{\overline{\alpha}} \cdot p_{kl}\Big\} \\ & \quad \cdot \Big\{[\sup(\nu_j)_{\overline{\alpha}} - \inf(\nu_j)_{\overline{\alpha}}] \cdot p_{ij} \cdot t + \sum_{(k,l)\in T_1\cup T_3} \sup(\nu_l)_{\overline{\alpha}} \cdot p_{kl} + \sum_{(k,l)\in T_2\cup T_4} \inf(\nu_l)_{\overline{\alpha}} \cdot p_{kl}\Big\}. \end{aligned}$$

$$g_{\alpha,p_{11},\dots,p_{mn}}[T_1,T_2,T_3,T_4,(i,j)](t) \stackrel{d}{=}$$
$$[\sup(\mu_i)_{\bar\alpha}\sup(\nu_j)_{\bar\alpha} - \inf(\mu_i)_{\bar\alpha}\inf(\nu_j)_{\bar\alpha}]\cdot p_{ij}\cdot t$$
$$+\sum_{(k,l)\in T_1}\sup(\mu_k)_{\bar\alpha}\sup(\nu_l)_{\bar\alpha}p_{kl} + \sum_{(k,l)\in T_2}\sup(\mu_k)_{\bar\alpha}\inf(\nu_l)_{\bar\alpha}p_{kl}$$
$$+\sum_{(k,l)\in T_3}\inf(\mu_k)_{\bar\alpha}\sup(\nu_l)_{\bar\alpha}p_{kl} + \sum_{(k,l)\in T_4}\inf(\mu_k)_{\bar\alpha}\inf(\nu_l)_{\bar\alpha}p_{kl}$$
$$-\{[\sup(\mu_i)_{\bar\alpha} - \inf(\mu_i)_{\bar\alpha}]\cdot p_{ij}\cdot t + \sum_{(k,l)\in T_1\cup T_2}\sup(\mu_k)_{\bar\alpha}\cdot p_{kl} + \sum_{(k,l)\in T_3\cup T_4}\inf(\mu_k)_{\bar\alpha}\cdot p_{kl}\}$$
$$\cdot\{[\sup(\nu_j)_{\bar\alpha} - \inf(\nu_j)_{\bar\alpha}]\cdot p_{ij}\cdot t + \sum_{(k,l)\in T_1\cup T_3}\sup(\nu_l)_{\bar\alpha}\cdot p_{kl} + \sum_{(k,l)\in T_2\cup T_4}\inf(\nu_l)_{\bar\alpha}\cdot p_{kl}\}.$$

Define for $\alpha \in (0,1)$ and $(p_{11},\dots,p_{mn}) \in \Pi_{m,n}$:

$$F_\alpha[p_{11},\dots,p_{mn}] \stackrel{d}{=}$$
$$\min\left\{ f_{\alpha,p_{11},\dots,p_{mn}}[T_1,T_2,T_3,T_4,(i,j)](t) \;\middle|\; \begin{array}{c} T_1+T_2+T_3+T_4 = \\ \{1,\dots,m\}\times\{1,\dots,n\}, \\ t\in[0,1],(i,j)\in T_2 \end{array} \right\}$$
$$G_\alpha[p_{11},\dots,p_{mn}] \stackrel{d}{=}$$
$$\max\left\{ g_{\alpha,p_{11},\dots,p_{mn}}[T_1,T_2,T_3,T_4,(i,j)](t) \;\middle|\; \begin{array}{c} T_1+T_2+T_3+T_4 = \\ \{1,\dots,m\}\times\{1,\dots,n\}, \\ t\in[0,1],(i,j)\in T_4 \end{array} \right\}$$

Then

$$\left\{ \begin{array}{l} \left[F_\alpha[P(\Gamma_1\cap\Sigma_1),\dots,P(\Gamma_m\cap\Sigma_n)],\right. \\ \left. G_\alpha[P(\Gamma_1\cap\Sigma_1),\dots,P(\Gamma_m\cap\Sigma_n)]\right] \end{array} \;\middle|\; \alpha\in(0,1) \right\}$$

is a set representation of $\mathrm{cov}(X,Y)$.

4. A limit theorem for estimating the covariance

In the last chapter we assumed that the probabilities $P(x=\mu_i$ and $Y=\nu_j)$ are known for $(i,j)\in\{1,\dots,m\}\times\{1,\dots,n\}$. What to do if we do not know these numbers ? We take a random sample, say of size N, and obtain N pairs of realizations of the fuzzy random vector (X,Y). By this sample, we can calculate the relative frequencies of all pairs $\mu_i,\nu_j)$ for $(i,j)\in\{1,\dots,m\}\times\{1,\dots,n\}$. They are estimators for the unknown probabilities. By using this estimators instead of the true values, we obtain an estimator for $\mathrm{cov}(X,Y)$.

We want to know what properties this estimator has. We define a metric on $Q(I\!R)$ in order to measure the distance between an estimator for the covariance and the covariance itself. The main result of this chapter is that the estimator converges almost sure against the covariance with respect to this metric.

We start with defining a metric on $Q(\mathbb{R})$ by transfering the Hausdorff pseudometric on $\wp(\mathbb{R})$.

Definition 4.1:
On the set of all nonempty subsets of $\mathbb{R}$ a pseudometric is given by

$$d_H(A,B) = \max\{\sup_{a\in A}\inf_{b\in B} |a-b|, \sup_{b\in B}\inf_{a\in A} |a-b|\}$$

for $A \subseteq \mathbb{R}$, for $B \subseteq \mathbb{R}$, $B \neq \emptyset$, $C \neq \emptyset$.

d_H is called the *Hausdorff pseudometric* (compare Artstein & Vitale [1]). On the set of all compact subsets of $\mathbb{R}$ d_H even is a metric.

With the help of this metric we define a metric on $Q(\mathbb{R})$:

Definition 4.2:
On $Q(\mathbb{R})$ a metric is given by

$$d_\infty(\mu,\nu) \stackrel{d}{=} \sup_{\alpha\in(0,1)} d_H[\mu_{\overline{\alpha}}, \nu_{\overline{\alpha}}] \qquad \text{for } (\mu,\nu) \in [Q(\mathbb{R})]^2.$$

If we replace the level sets by any other set representation, the distance between two fuzzy sets remains the same.

In a natural way we obtain a convergence notion on $Q(\mathbb{R})$:

Definition 4.3:
Let $\{\mu_n\}_{n\in\mathbb{N}}$ be a sequent of elements of $Q(\mathbb{R})$ and $\mu \in Q(\mathbb{R})$
$\{\mu_n\}_{n\in\mathbb{N}}$ is called *Hausdorff convergent against*
— in signs: $\{\mu_n\}_{n\in\mathbb{N}} \xrightarrow{d_\infty} \mu$ —
if and only if $\lim_{n\to\infty} d_\infty(\mu_n,\mu) = 0$.

Our task is to estimate the unknown covariance of X and Y. In order to get sensitive estimators for the unknown probabilities $P(x = \mu_i \text{ and } Y = \nu_j)$ for $(i,j) \in \{1,\ldots,m\} \times \{1,\ldots,n\}$, we have to demand that the conditions for taking the sample remain constant, i.e. that we deal with an i.i.d.-sequence. We define for the fuzzy case:

Definition 4.4:
Let $(X_N, Y_N) : \Omega \to [Q(\mathbb{R})]^2$ be a fuzzy random vector for all $N \in \mathbb{N}$, and let $\{\mu_1,\ldots,\mu_m\} \times \{\nu_1,\ldots,\nu_n\}$ be a subset of $Q[(\mathbb{R})]^2$.
$\{X_N, Y_N)\}_{N\in\mathbb{N}}$ is called *i.i.d.-sequence* with the codomain $\{\mu_1,\ldots,\mu_m\} \times \{\nu_1,\ldots,\nu_n\}$ if and only if

(i) $\{\mu_1,\ldots,\mu_m\} \times \{\nu_1,\ldots,\nu_n\}$ is the codomain of (X_N,Y_N) for all $N \in \mathbb{N}$

(ii) $\{(U_N, V_N) \mid \in \mathbb{N}\}$ is a (usual) i.i.d.-sequence of random vectors where

$$U_N(\omega) = i \overset{\mathrm{d}}{\Longleftrightarrow} X_N(\omega) = \mu_i$$

and

$$V_N(\omega) = j \overset{\mathrm{d}}{\Longleftrightarrow} Y_N(\omega) = \nu_j$$

Now we can formulate the main result of this chapter:

Theorem 4.5:
Let $\{(X_N, Y_N)\}_{n \in \mathbb{N}}$ be an i.i.d.-sequence of finite fuzzy random vectors on $(\Omega, \mathcal{A}, P)$. Let $\{\mu_1, \ldots, \mu_m\} \times \{\nu_1, \ldots, \nu_n\} \subseteq [Q(\mathbb{R})]^2$ contain the common codomain of $\{(X_N, Y_N)\}_{n \in \mathbb{N}}$. Define for $\omega \in \Omega$, $(i,j) \in \{1, \ldots, m\} \times \{1, \ldots, n\}$, $N \in \mathbb{N}$:

$$p_{ij}^{(N)}(\omega) \overset{\mathrm{d}}{=} \frac{1}{N} \cdot \mathrm{card}\{k \in \{1, \ldots, N\} \mid X_k(\omega) = \mu_i \text{ and } Y_k(\omega) = \nu_i\}$$

$p_{ij}^{(N)}$ is the relative frequency of (μ_i, ν_j) in the random sample. Define for $\omega \in \Omega$, $n \in \mathbb{N}$ the estimator $B_\omega^{(N)} \in Q(\mathbb{R})$ by

$$B_\omega^{(N)} \overset{\mathrm{d}}{=} \sup\left\{\alpha \cdot I_{\left[F_\alpha[p_{11}^{(N)}(\omega), \ldots, p_{mn}^{(N)}(\omega)], G_\alpha[p_{11}^{(N)}(\omega), \ldots, p_{mn}^{(N)}(\omega)]\right]}(t) \mid \alpha \in (0,1)\right\}.$$

Then there exists a zero set $M \in \mathcal{A}$ such that

$$\{B_\omega^{(N)}\} \xrightarrow{d_\infty} \mathrm{cov}(X, Y)$$

for $\omega \in \Omega \setminus M$.

Remark The practical calculations with fuzzy sets turned out to be intricate. Therefore a software tool was developed by Kruse [3] which supports a user in working with fuzzy data.

5. References

1. Artstein, Z. and Vitale, R.A. (1975) 'A strong law of large numbers of random compact sets', *Ann. Probability 3*, 879–882.
2. Kruse, R. (1987) 'On the variance of random sets', *J. Math. Anal. Appl. 122*, 469–473.
3. Kruse, R. (1987) 'On a software tool for statistics with vague data', *Fuzzy sets and systems 24*, 377–383.

4. Kruse, R. and Meyer, K.D. (1987) 'Statistics with vague data', *D. Reidel Publ. Comp., Dordrecht and Boston, Monography.*
5. Kwakernaak, H. (1977) 'Fuzzy random variables', Part I: 'Definitions and theorems', *Inform. Sci. 15*, 1–15. Part II: 'Algorithms and examples for the discrete case', *Inform. Sci. 17*, 253–278.
6. Miyakoshi, M. and Shimbo, M. (1983) 'Set representations of a fuzzy set and its application to Jensen's inequality', *Submitted for publication.*
7. Puri, M.L. and Ralescu, D.A. (1986) 'Fuzzy random variables', *J. Math. Anal. Appl. 114*, 409–422.
8. Zadeh, L.A. (1965) 'Fuzzy sets', *Information and control 8*, 338–353.
9. Zadeh, L.A. (1975) 'The Concept of a Linguistic Variable and its Application to Approximate Reasoning', Part I: *Inform. Sci. 8*, 199–249. Part II: *Inform. Sci. 8*, 301–357. Part III: *Inform. Sci. 9*, 43–80.

FUZZY MEASURES WITH DIFFERENT LEVELS OF GRANULARITY

L. M. DE CAMPOS, S. MORAL
Departamento de Ciencias de la Computación
e Inteligencia Artificial
Universidad de Granada
18071-Granada
Spain

ABSTRACT. In this paper we will represent the uncertainty about a particular event, A, by means of an interval [l(A),u(A)]. The extremes of such interval are linguistic values from a particular scale or level of granularity. The problem is how to transform an uncertainty value or interval with arbitrary extremes into an interval expressed on a specific scale. Two conditions are requested to do this transformation: i) Improper information, not contained in the original, must not be obtained, ii) The loss of information must not be too great. Several approaches are considered, and their properties and behaviour are studied. Most of them are based on different methods of ranking fuzzy numbers. Finally, several examples of particular transformations are considered.

1. Introduction

In the modeling of complex systems, it becomes necessary to consider knowledge that involves some uncertainty. However, this information may hardly be represented using an only approach. In most of the cases several formalisms or mixed ones have to be used. In this paper we consider the situation in which the uncertainty about the values of a variable is probabilistic in essence, but our knowledge about the exact probabilities is fuzzy.

Let us suppose that we have a variable, Y, taking its values on a set X. If a probabilistic information, P, is available about Y, then P(A) may be interpreted as the degree of confidence or certainty about the truth of the proposition 'Y belongs to A'. More precisely, we consider this degree of certainty as the limit of the relative frequencies of the verification of the event 'Y belongs to A', in the present conditions.

However the values of probability suppose a perfect knowledge about the way how Y takes its values, the exact frequency of elements, w, from Ω, in which $Y(w) \in A$, and this limit is not always known.

In the cases in which our knowledge is not precise enough to be able to determine exact probabilities,we may establish two bounds: a lower bound, l(A), an upper bound, u(A), and its corresponding interval [l(A),u(A)] (see Dempster (1967), Shafer (1976)). But if we do not have

W. H. Janko et al. (eds.), Progress in Fuzzy Sets and Systems, 134–146.

a numerical source of information, the assignation of real values from the interval zero-one to the bounds of such interval may be very difficult. Moreover, if we get the information from an expert, he can be very reluctant to express its degree of uncertainty about an event, using real numbers.

In this case, it seems more appropriate to use a linguistic scale, in the sense of Bonissone and Decker (1986). In this way, we may say 'This event is very likely', 'The probability of such event is between 0.6 and 0.8, approximately', etc.

A linguistic scale is also called a level of granularity that, according to Bonissone and Decker (1986), may be interpreted as 'the finest level of distinction among different quantifications of uncertainty, that adequately represents the user's discriminant perception'.

The problem that is considered in this paper is the following: we have a working level of granularity, D, and we have information expressed by means of fuzzy subsets that are not in this scale. Then it is necessary to transform this fuzzy value or interval into another interval with extremes in D. This situation may happen, for example, if we want to combine informations defined on different levels, to codify the result on an appropriate scale, or to codify informations stated on an arbitrary way, etc.

An important point is that the resulting interval must have approximately the same information, the same meaning that the original one. We have split this requirement in two separate aspects:

i) We can not obtain improper information, not contained in the original.

ii) The loss of information must not be great.

The importance of these conditions is not the same. The former will be an essential restriction, but the latter will depend on the objective scale, D. If the linguistic values of D are very vague, then it will be difficult to codify precise statements without loosing any information. Bonissone's approach (Bonissone (1979)) to this problem is based on the use of two parameters: the first moment of the distribution and the area under the curve. Then a weighted Euclidean distance is used to determine the term in D with the most similar meaning to the initial one. Our approach differs in the use of intervals to represent the uncertainty values. This allows to consider the incompleteness of the information and a more close representation, at the same level of granularity. The use of intervals allows us to use $n(n+1)/2$ different expressions from a scale with n linguistic labels.

2. Preliminary Definitions

2.1. THE AVERAGE VALUE

First we consider the definition of fuzzy number from Campos and González (1989).

Definition. A fuzzy subset A, of the real line, with membership function

$\mu_A(\cdot)$ is said to be a fuzzy number iff:

i) $\forall\alpha\in[0,1]$, $A_\alpha=\{x\in\mathbb{R} \;/\; \mu_A(x)\geq\alpha\}$ (α-cut of A) is a convex set.

ii) $\mu_A(\cdot)$ is an upper semicontinuous function.

iii) A is normalized, i.e., $\exists m\in\mathbb{R} \;/\; \mu_A(m)=1$.

iv) $\mathrm{Supp}(A) = \{x\in\mathbb{R} \;/\; \mu_A(x)>0\}$ is a bounded set of $\mathbb{R}$.

Dubois and Prade (1980) defined fuzzy numbers as unimodal fuzzy numbers. This condition is omitted in the definition above.

The following definition (Campos and González (1989)) uses the parametric function:

$$f_A^\lambda(\alpha) = \begin{cases} \lambda b_\alpha+(1-\lambda)a_\alpha & \text{if } 0\leq\alpha\leq1 \\ 0 & \text{otherwise} \end{cases}$$

with $A_\alpha=[a_\alpha,b_\alpha]$.

Definition. The average value of a fuzzy number A with respect to an additive measure S on [0,1] and $\lambda\in[0,1]$ is the value

$$V_S^\lambda(A) = \int_0^1 f_A^\lambda(\alpha)\, dS(\alpha)$$

The parameter λ is selected to choose a single position for each interval $A_\alpha=[a_\alpha,b_\alpha]$. The average value represents a mean value of the different α-cut positions through a measure S on [0,1].

When λ takes every value in [0,1] we obtain an interval of average values, limited by V_S^0 and V_S^1:

$$[V_S^0(A),V_S^1(A)] = \{V_S^\lambda(A) \;/\; \lambda\in[0,1]\}$$

If the Lebesgue measure L is considered, this interval coincides with the mean value of a fuzzy number defined by Dubois and Prade (1987):

$$E(A) = [E_*(A),E^*(A)] = [V_L^0(A),V_L^1(A)],$$

where $E_*(A)$, $E^*(A)$ are the Choquet integrals (Choquet (1953/54)):

$$E_*(A) = \int x\, dN, \qquad E^*(A) = \int x\, d\Pi,$$

and Π, N are the possibility-necessity measures associated with the fuzzy number A (see Zadeh (1978)). $E_*(A)$ and $E^*(A)$ are called the lower and upper mean values of A, respectively. The average value under the Lebesgue measure may be expressed as:

$$V_L^{\lambda}(A) = \lambda E_*(A)+(1-\lambda)E^*(A)$$

An immediate modification of these values may be achieved by using the Sugeno integral instead of the Choquet integral. Then we obtain the Sugeno mean values:

$$T_*(A) = \int x \circ N, \quad T^*(A) = \int x \circ \Pi$$

$T^*(A)$ was introduced by Yager (1981) as a ranking function for fuzzy numbers on the unit interval. $T_*(A)$ was considered by González (1989). The two values define the Sugeno mean interval

$$ES(A) = [T_*(A), T^*(A)]$$

2.2. LEVEL OF GRANULARITY

A linguistic scale or level of granularity is a classification of the interval [0,1] in a finite class of fuzzy subsets

$$D = \{C_0, C_1, C_2, .., C_n\}$$

where a linguistic label is assigned to every subset. In order to be possible to define fuzzy measures taking values in D, the following regularity conditions are considered:

i) $C_0=\{0\}$, $C_n=\{1\}$.

ii) P is totally ordered: $C_0 \leq C_1 \leq ... \leq C_{n-1} \leq C_n$.

iii) P has a symmetry property: $1-C_i=C_{n-i}$, $i=0,..,n$.

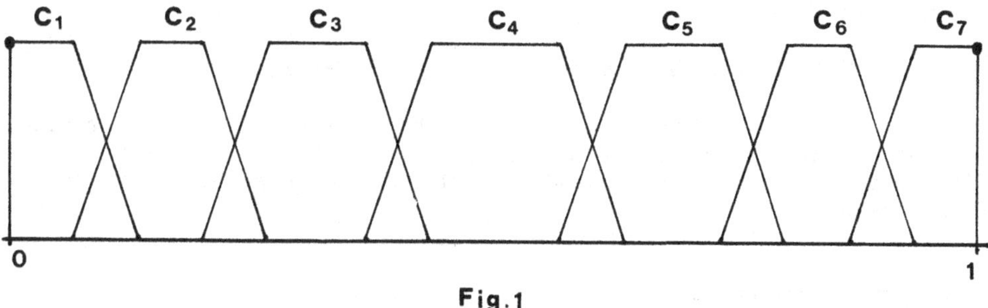

Fig.1

Example. We may consider the linguistic scale $D = \{C_0, C_1, .., C_6\}$, with fuzzy subsets as in Fig.1 and the following linguistic labels:

C_0: Impossible
C_1: Extremely unlikely
C_2: Very low chance
C_3: Small chance
C_4: It may
C_5: Meaningful chance
C_6: Most likely
C_7: Extremely likely
C_8: Certain

2.3. PAIRS OF FUZZY MEASURES

We use the formalism of fuzzy measures to represent the uncertainty about the values of a variable, more concretely the concept of an ordered pair of fuzzy measures.

Definition. Let X be a finite set and D a level of granularity. (l,u) is said to be a pair of ordered fuzzy measures if and only if l and u are mappings from $P(X)$ to D such that

1) $l(\emptyset) = u(\emptyset) = C_0$.
2) $A \subseteq B \Rightarrow l(A) \leq l(B),\ u(A) \leq u(B)$.
3) $l(A) = 1 - u(\bar{A})$.

For each $A \subseteq X$ we have two values of certainty: the lower value, l(A), and the upper value, u(A). Therefore a certainty interval [l(A),u(A)] can be associated with A. This interval may be considered as our lack of precision in determining an exact value of certainty. For example, in the case of complete ignorance, this interval is maximum $[C_0, C_n]$. When our knowledge is more specific, then this interval is reduced. In this sense, the following definition of inclusion may be given.

Definition. If (l,u) and (l',u') are pairs of ordered fuzzy measures defined on levels D and D' respectively, then (l',u') is said to be included in (l,u) if and only if

$$\forall A \subseteq X \ \ l'(A) \leq l(A) \ \text{(or equivalently } u'(A) \geq u(A)\text{)}$$

Remark: If levels D and D' are equal, then there is not any problem, because of the total order property of D, but when D and D' are different, the inclusion depends on the selected method for ranking fuzzy numbers.

3. Ranking Fuzzy Numbers

In the problem of transformation between different scales, we first require the condition of not obtaining improper information. Following the definition of inclusion of pair of ordered fuzzy measures, if we transform the pair (l,u) into the pair (l',u'), then (l',u') have to be included in (l,u). This restriction is verified when $l'(A) \leq l(A)$ (or

$u'(A) \geq u(A)$).

There is a total order defined among the fuzzy numbers from a specific scale, but $l'(A)$ and $l(A)$ are from different scales; then we must consider general methods of ranking fuzzy numbers in order to test whether (l', u') is included in (l, u).

In the same way, the second condition may be related with the problem of ordering fuzzy numbers.

The methods we have considered more relevant to our problem are the following:

a) The Dubois and Prade's definition of maximum. Starting from the definition of fuzzy maximum (Dubois and Prade (1980)), based on the Zadeh's extension principle (see Fig.2)

$$\tilde{Max}(B,C) = D,$$

where

$$\mu_D(x) = \operatorname*{Max}_{y,z \in \mathbb{R}} \{ \min(\mu_B(y), \mu_C(z)) \;/\; x = \max(y,z) \}$$

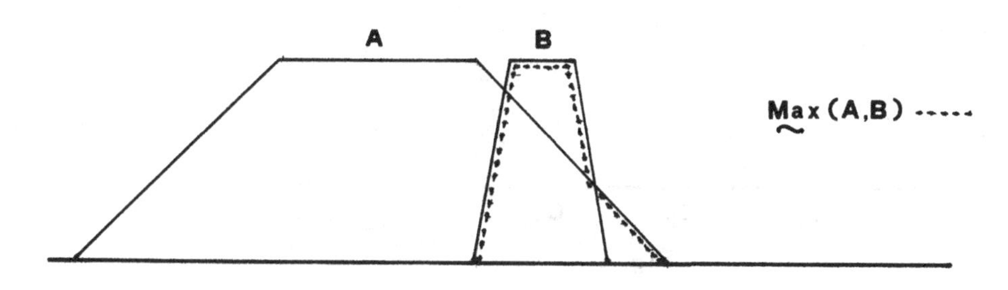

Fig.2

b) The González's approach (González (1989)) based on the choice of an interval order relation to compare the intervals defined by an appropriate mean value for each fuzzy number, that is,

1. To define $M(A) = [m_*(A), m^*(A)]$, mean value of A.
2. To define an interval order relation $\leq$.

Then, an order relation may be defined for fuzzy numbers by means of:

$$B \leq C \iff [m_*(B), m(B)] \leq [m_*(C), m^*(C)]$$

Next we give some examples of order relations for intervals:

1) $[a,b] \leq [c,d]$ iff $b \leq c$.
2) $[a,b] \leq [c,d]$ iff $a \leq c$.
3) $[a,b] \leq [c,d]$ iff $b \leq d$.
4) $[a,b] \leq [c,d]$ iff $\lambda a + (1-\lambda)b \leq \lambda c + (1-\lambda)d$, where $\lambda \in [0,1]$.

4. Transformation of linguistic values

We shall only consider how to transform the lower bound l(A) of the uncertainty interval [l(A),u(A)]. The upper bound may be transformed by duality: First, we obtain $l(\bar{A})=1-u(A)$, then $l(\bar{A})$ is codified as $l'(\bar{A})$, and the transformation of u(A), $u'(A)=1-l'(\bar{A})$, is obtained.

The problem is stated in the following way: We have a granularity level, $D = \{C_0, C_1, .., C_n\}$ and a value l(A)=E, a fuzzy number not belonging to D. Then, an element from D, C_k, must be selected in such a way that (see Fig.3)

a) we do not obtain improper information, that is $C_k \leq E$

b) The loss of information, i.e., the difference $E-C_k$ is not great.

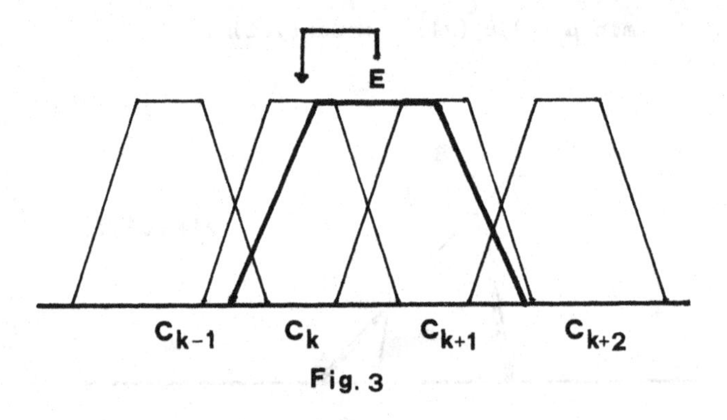

Fig. 3

4.1. METHODS BASED ON A COMPATIBILITY RELATION

A measure of compatibility between two fuzzy numbers is established. For example, we may consider (see Fig.4)

$$Co(C,E) = \underset{x\in[0,1]}{\text{Max}} \{ \min(\mu_C(x),\mu_D(x)) \}$$

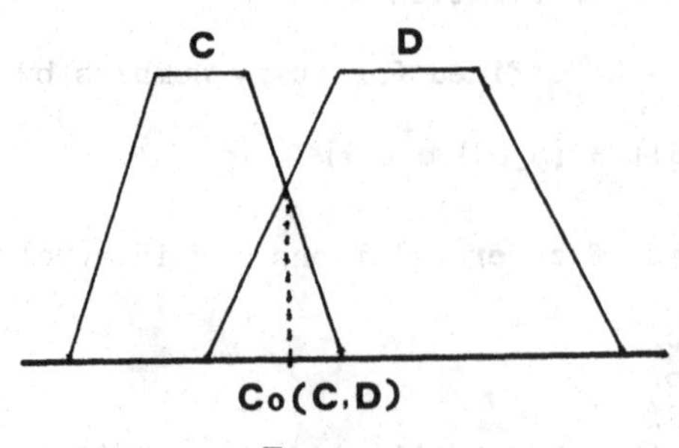

Fig. 4

4.1.1. The first method transforms $l(A)=E$ on the value C_k such that the compatibility between E and C_k is maximum:

$$Co(E,C_k) = \underset{C_j \in D}{\text{Max}} \{ Co(E,C_j) \}$$

This method is similar to the linguistic approximation from Bonissone (1979). They consider a compatibility relation based on a weighted Euclidean measure of the first moment of the distribution and the area under the curve.

This approach does not take into account that $l(A)=E$ is a lower bound of our uncertainty interval, and because of this in many situations we obtain improper information. For example, in Fig. 5, E is transformed on C_k ($Co(E,C_k)=1$), and the lower bound has not been decreased.

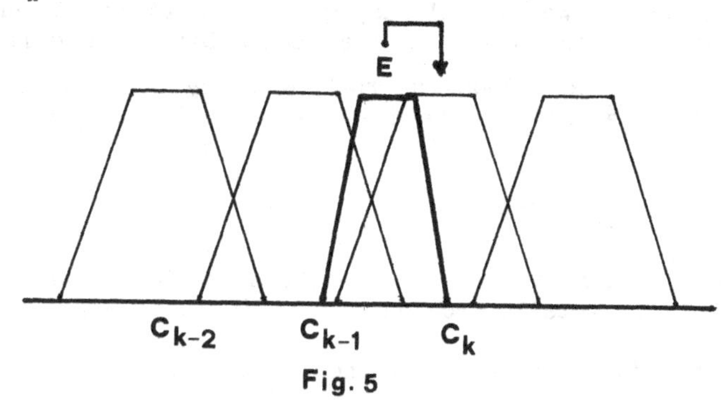

Fig. 5

4.1.2. To solve the problem of obtaining improper information, a crisp compatibility relation is determined: C and E are said to be compatible if and only if $Co(C,E)>0$. Then $l(A)=E$ is codified as

$$C_k = \text{Min} \{ C_j \;/\; Co(E,C_j)>0 \}$$

In this way, we do not obtain improper information; however, the loss of information may be very large (see Fig. 6).

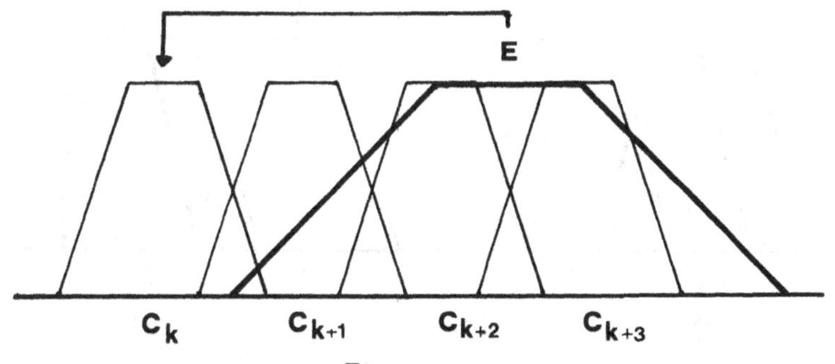

Fig. 6

4.1.3. An intermediate solution between *4.1.1.* and *4.1.2.* may be achieved by considering that C and E are compatible if and only if their compatibility is greater than a parameter, ε, belonging to [0,1]:

$$Co(C,E) > \varepsilon$$

When ε=0, this method coincides with *4.1.2.* and when ε is very close to one, is very similar to *4.1.1.* We may avoid the problems of the above methods by taking an intermediate value. For example, taking ε=0.25, in Fig.5 we obtain the transformation $E \longrightarrow C_{k-1}$ and, in Fig.6, $E \longrightarrow C_{k+1}$.

4.2. METHODS BASED ON AN ORDER RELATION ON FUZZY NUMBERS

This approach consists on determining a criterion for not obtaining improper information. As we have already said, this is equivalent to the condition: if we transform $E \longrightarrow C_k$, then $E \geq C_k$. Different criteria are obtained depending on the method for ranking fuzzy numbers considered. Once a particular method has been chosen, the fuzzy value E is transformed in

$$C_k = \text{Max} \{ C_i \ / \ C_i \leq E \}$$

This maximum does not have any problem because the elements from the linguistic scale D are totally ordered.

4.2.1. The first method is based on the order from Dubois and Prade (1980):

$$E \geq C_k \Leftrightarrow \underset{\sim}{\text{Max}} (E, C_k) = E$$

The resulting transformation may be very sensitive to small variations of the shapes of the membership functions in points of low membership degree (see Fig.7: In a) we obtain $E \longrightarrow C_k$, whereas in b) $E \longrightarrow C_{k-1}$ is obtained).

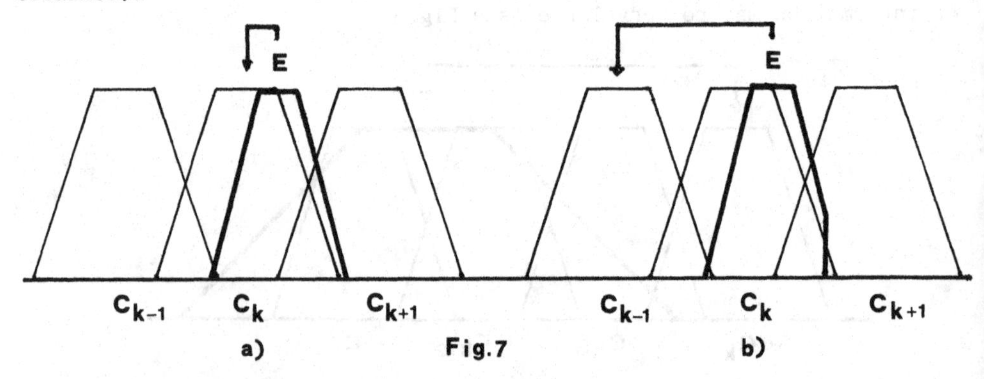

Fig.7

4.2.2. Following the González's approach, an order relation for fuzzy numbers may be determined by considering a mean interval for fuzzy numbers and an interval order relation. In this case, we consider the Sugeno mean value:

$$ES(E) = [T_*(E), T^*(E)]$$

and the interval order

$$[a,b] \leq [c,d] \Leftrightarrow b \leq c.$$

It can be easily obtained that

$$T_*(E) = \underset{x\in[0,1]}{\text{Max}} \{\min(x, \mu_E(x))\}, \quad T^*(E) = \underset{x\in[0,1]}{\text{Max}} \{\min(x, 1-\mu_E(x))\}$$

These values are close to the peak or flat of the fuzzy number E, when E is near to zero, and close to the extremes of E when E is near to one (see Fig. 8).

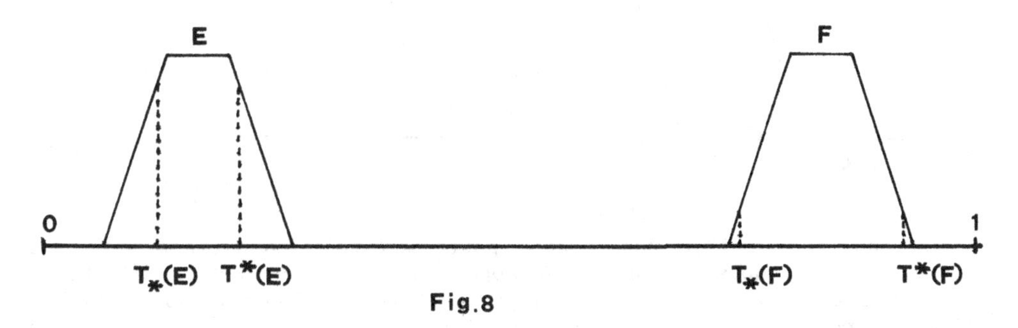

Fig.8

The transformation of E is equal to C_k, where

$$C_k = \text{Max}\ \{C_i \in D \ /\ T^*(C_i) \leq T_*(E)\}.$$

When we transform values close to one, we obtain a similar behaviour to method *4.1.2.*. Thus, in this case, we have loss of information.

4.2.3. If the Sugeno integral is replaced by the Choquet integral in the calculation of the mean value of a fuzzy number:

$$EC(A) = [E_*(A), E^*(A)],$$

then the transformation is more regular along the unit interval. The relative position of $E_*(A)$, $E^*(A)$ with respect to A are independent of the position of A on [0,1], because of the following well-known property of the Choquet integral:

$$E_*(a+bA) = a+bE_*(A), \quad E^*(a+bA) = a+bE^*(A), \quad \forall a\in\mathbb{R}, \ \forall b\in\mathbb{R}^+.$$

4.2.4. Finally, the last method we propose takes into account that the fuzzy quantity we want to transform, E, is the lower bound of the interval [l(A),u(A)]. It uses the Choquet integral (as *4.2.3*) but it assumes that only the left part of E (the part determining the lower limit) is relevant. For example, by method *4.2.3.* (see Fig.9) we have $E\longrightarrow C_{k-1}$. Nevertheless, we may assume that the transformation $E\longrightarrow C_k$ does not obtain improper information, considering the left parts. This is equivalent to consider the interval order:

$$[a,b] \le [c,d] \Leftrightarrow a \le c.$$

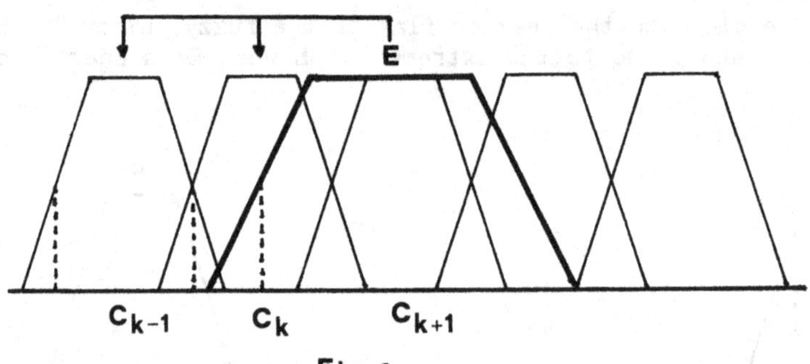

Fig.9

The resulting method has a good performance in comparison with the former ones. It does not obtain improper information, as *4.1.1.* It has a smaller loss of information than *4.1.2*, *4.2.1*, *4.2.2* and *4.2.3.* Finally, it has a more sound theoretical ground that *4.1.3.* We think that it is appropriate to define general transformations, showing a good balance between the two conditions we have imposed. A similar consideration about the lower limits can be applied to methods *4.2.1 and 4.2.2*, improving their performance.

5. Conclusions

In this paper we have studied the codification of uncertainty values, l(A) and u(A), which have been expressed as fuzzy numbers on the unit interval on a determined linguistic scale $D = \{C_0, C_1, .., C_n\}$. Several methods have been proposed and their performance analyzed. The transformation we consider as the most appropriate is the following:

$$l(A) = E \longrightarrow \text{Max } \{C_i \in D \ / \ E_*(C_i) \le E_*(E)\}$$

$$u(A) = F \longrightarrow \text{Min } \{C_i \in D \ / \ E^*(C_i) \ge E^*(F)\}$$

There are some particular cases where another method could be more suitable. For example, if we are transforming fuzzy values with very different shapes, the Choquet integral, as a mean value, may produce non easily interpetable compensations. Method *4.1.3* could be appropriate in these situations.

Next, we give some examples from scale D (Fig.1) to scale D'={impossible, Approx. 0, Approx. 0.1,..., Approx. 0.9, Approx. 1, 1}, where approximately 0.i is represented as in Fig.10 and the metod used is *4.2.4*.

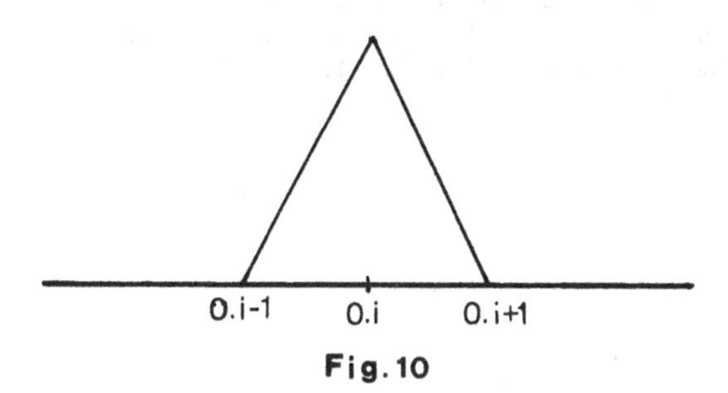

Fig.10

-The interval [It may, Most likely] is transformed on [Approx. 0.4, Approx. 0.9].

-The value Small chance (identified with the interval [small chance, small chance]) is transformed on [Approx. 0.2, Approx. 0.4].

-Extremely unlikely is transformed on [Impossible, Approx. 0.1].

Examples of the inverse transformations are:

-[Approx. 0.4, Approx. 0.6] ⟶ [Small chance, Meaningful chance].

-Approx. 0.7 ⟶ Meaningful chance.

References

Bonissone, P.P. (1979) 'A pattern recognition approach to the problem of linguistic approximation in System Analysis', Proceedings of the IEEE International Conference on Cybernetics and Society, Denver, 793-798.

Bonissone, P.P. and Decker, K.S. (1986) 'Selecting uncertainty calculi and granularity :an experiment in trading-off precision and complexity, in L.N. Kanal, J.F. Lemmer, (eds), Uncertainty in Artificial Intelligence, North-Holland, 217-247.

Campos, L.M.de and González, A (1989) 'A subjective approach for ranking fuzzy numbers', Fuzzy Sets and Systems 29, 145-153.

Choquet, G. (1953/54) 'Theory of capacities', Ann. Inst. Fourier 5, 131-292.

Dempster, A.P. (1967) 'Upper and lower probabilities induced by a multivaluated mapping', Ann. Math. Statist. 38, 325-339.

Dubois, D. and Prade, H. (1980) Fuzzy Sets and Systems. Theory and Applications, Academic Press, New York.

Dubois, D. and Prade, H. (1987) 'The mean value of a fuzzy number', Fuzzy Sets and Systems 24, 279-300.

González, A. (1989) 'A study of the ranking function approach through mean values', To appear in Fuzzy Stes and Systems.

Shafer, G. (1976) A Mathematical Theory of Evidence, Princeton University Press, Princeton.

Yager, R.R. (1981) 'A procedure for ranking fuzzy subsets of the unit interval', Information Sciences 24, 143-161.

Zadeh, L.A. (1978) 'Fuzzy sets as a basis for a theory of possibility', Fuzzy Sets and Systems 1, 3-28.

GRAPH-BASED FORMULATION OF FUZZY DATA

Yasuto Shirai, Issei Fujishiro† and Tosiyasu L. Kunii
Department of Information Science
Faculty of Science
The University of Tokyo
7-3-1 Hongo, Bunkyo-ku
Tokyo 113 Japan

ABSTRACT: It has been pointed out that a database capable of handling fuzzy data will extend the frontier of the database application domain. Formulation of fuzzy data based on the relational theory has been reported in the literature, but such formulation suffers from the model's own characteristics, namely, the atomicity imposed on each data item and the uniformity required of the data organization. This paper proposes a new way of modeling fuzzy data using the framework of Extended Graph Data Model (EGDM). EGDM is a post-relational link-oriented data model, and has stronger modeling capability than the relational model. Nature of fuzzy data is investigated from the perspective of data modeling, and a fuzzy property of an entity is modeled using the EGDM constructs. A restricted class of the entity-relationship model is used in the intermediate stage of this modeling process. An example is used to illustrate the features of our formulation and to demonstrate its operational aspect.

1. Introduction

A database is a formulation of a human perception on a part of the real world that is of interest to the enterprise involved. Though a database is to be manipulated through the use of a computer, the object modeled by a database exists not in the machine but in the real world. It is the human perception that bridges the gap between the real world and the machine. The very use of human perception, however, leads to several problems.

One problem is rooted in the difference in the degree of precision employed by the human perception and by the machine operation. The human beings make many imprecise, qualitative judgements; statements such as "He is *very tall*," and "Linda has *beautiful* eyes," are qualitative and somewhat subjective. Though their meaning is clear to us, there is no way for a computer to manipulate such statements directly. On the other hand, a precise statement such as "His height is *198.5*cm," can be readily handled by a computer, while we tend to avoid this type of statement in our daily conversation.

---

†Current address: The Institute of Information Sciences and Electronics, University of Tsukuba, 1-1-1 Tenno-Dai, Tsukuba City, Ibaraki 305 Japan

W. H. Janko et al. (eds.), Progress in Fuzzy Sets and Systems, 147–160.

The fuzzy set theory provides a foundation for the manipulation of imprecise information. A fuzzy set is capable of expressing an imprecise and/or subjective concept of a human being in a well-defined manner. The cases where fuzzy data may be useful are [22]:

(1) when exact values are not known, but some knowledge of possible value distribution is available;

(2) when the data is inherently fuzzy; and

(3) when the data is qualitative, rather than quantitative.

It has been argued that a database with a capability for handling fuzzy data (i.e., a *fuzzy database*) will contribute greatly to a variety of fields, such as knowledge base systems and expert systems [10].

The ultimate goal of our research is to construct a fuzzy database, and this paper proposes a new approach to the modeling of fuzzy data as the first step toward that goal. Clearly, it is very much desired that our formulation be transformable into the framework of some data model. For this reason, we have chosen a weighted di-graph as the basis for our formulation of fuzzy data. A weighted di-graph is also a basis of a link-oriented data model called *Extended Graph Data Model* (EGDM). Because of a common structural basis, the transformation of our formulation into an EGDM schema reduces to a straightforward task.

Our approach is different from the approaches adopted in other similar attempts in one fundamental aspect. In other approaches, the structural organization of a conventional data model, in most (if not all) case, the relational model, is so extended as to accommodate fuzzy sets within the framework of the extended model. This would also require an extension of the manipulation facility of the model. In out research, however, the nature of the fuzzy data is investigated more closely, and modeled using the constructs available in a conventional data model, namely, those of EGDM. Such approach would require no extension on the manipulation facility of the model.

Section 2 briefly surveys the development of fuzzy data models, and identifies their problems. Then in Section 3 we look more closely into the nature of fuzzy data from the perspective of data modeling, and propose a technique for modeling fuzzy data. Here we are interested in the membership function in particular. In modeling the membership function, we adopt the concept of the entity-relationship (E-R) model [2, 4, 5], and come up with a graph-based modeling framework for fuzzy data. In order to compensate for the lack of data manipulation facility in the E-R model, we project this technique onto the realm of EGDM (Section 4). EGDM has a data manipulation facility defined in the form of a data language called Graph Data Language, and, as far as the structural aspect is concerned, it is equivalent to a restricted class of the E-R model [8]. Section 5 illustrates our formulation with an example. Data manipulation facility of Extended Graph Data Model as applied to fuzzy data is demonstrated. Section 6 concludes the paper with some comments on future research directions.

2. Fuzzy Relational Data Models

2.1. DEVELOPMENT OF FUZZY RELATIONAL DATA MODELS

The theoretical study on data models did not gain much attention until the introduction of the relational model by Codd [6] in 1970. The basis of the relational theory is the set theory. Since the fuzzy set theory is in a sense a generalization of the classical set theory, the extended relational model capable of handling fuzzy data has attracted much attention among the researchers. The history of fuzzy databases overlaps with that of the *fuzzy relational databases*. In the following discussion, therefore, when we talk about a database, it is assumed that it is a 'relational' database.[1]

In the field of the database systems, the handling of fuzzy data has been studied in a wider scope of incomplete or uncertain information management. Grant [11, 12] was among the first to point out the issue of incomplete information in a database. The treatment of *null values*, in particular, has drawn much attention from the researchers. A good introduction on the subject is provided in [7]. This problem, however, is far from being settled and no satisfactory solution has been developed to cope with the general situation. It is even claimed that the whole topic is a matter of personal taste [17].

The subject of incomplete or partial information is encompassed by the issue of fuzzy data management. The term "fuzzy database" may be traced back to DATAPLAN [14]. In DATAPLAN, a database relation is extended to a fuzzy subset of a Cartesian product of domains; each tuple in a relation has a grade of membership associated with it indicating the degree with which the tuple belongs to the relation. It is a simple extension of the relational model, and can be readily implemented on any relational system by adding an extra column to each relation for holding the membership grade. Due to its simplicity, however, it has only a limited description capability.

Buckles and Petry [1] further extend the notion of a relation to a Cartesian product of power sets of domains,[2] and replace the usual equality relation on a domain with its fuzzified version, i.e., a *similarity relation*. The key aspect of their fuzzy relational database is "that domain values need not be atomic." Clearly, this is in conflict with the Codd's first normal form assumption. Also, as the result of introducing a similarity relation, we must define a set of relational operations for each such relation. This would degrade the high degree of uniformity, which is a major advantage of the relational model.

Umano [20] and Zemankova-Leech and Kandel [21] propose more general models for modeling fuzzy data, and implement prototype systems based on their respective models. They introduce the notion of possibility into relational databases, and a possibilistic distribution and a null value as well as an atomic value (e.g., an integer and a character string) are allowed as an attribute value.

2.2. TWO PROBLEMS

Though their fuzzy relational models are very rich in expressive power, they have their flaws. First of all, the explicit introduction of nulls necessitates a special treatment. In order to fully support nulls, one question must be answered: "Is a null equal to another null, or are they not?" As mentioned earlier in this section, this question itself has not been given any

---

[1]To the authors' knowledge, there has been only one reported case of a non-relational fuzzy database [22]. But the approach taken there is basically the same as that of the fuzzy relational data model. The entity-relationship model is extended to allow for fuzzy entity/relationship types and fuzzy attributes. When an E-R schema so constructed is translated into a logical schema, the same problems will arise.

[2]To be more precise, they first take a set of all non-empty subsets of each domain, and then define a database relation as a subset of a Cartesian product of these sets of non-empty sets. This procedure avoids the introduction of a null value as a domain value.

Name	# of children
Tom	3
John	{0.3/0, 0.8/1, 1.0/2}
Mike	?

Fig. 1 Relation-based representation.

satisfactory answer. The next issue to be considered is the assumption on domains. The relational model implicitly assumes that each domain is a set of atomic values. If we follow this assumption strictly, then the introduction of a possibilistic distribution, which is in itself a fuzzy set, as a domain value will clearly be a violation.

The two problems discussed above are illustrated in Fig. 1. Here the table maintains the information on the number of children each person has. There is no problem with Tom's case; he has exactly three children and we know it. For John, we only have a vague idea which may be described as ''John has two or less children. It could be that he has none.'' So we have a fuzzy set representing a possibilistic distribution as an attribute value for John's tuple. Finally, for Mike, the attribute ''# of children'' is given a null value. There are at least two ways to interpret this null value:

(1) we simply do not know how many children he has; or

(2) this property itself is not applicable to Mike (e.g., he may be just two years old).

This simple example makes it clear that the relation-based representation of fuzzy data behaves rather badly, and that the uniformity and the simplicity of the relational model is lost.

In the next section, we will describe how these problems can be avoided by using the graph structure for modeling fuzzy data.

3. Graph-based Representation of Fuzzy Data

3.1. DATABASE DESIGN PROCESS

In this section, we quickly go over the logical database design process based on the entity-relationship (E-R) approach. For more details on this approach, see, for example, [3, 5].

As the first step in the database design process, *elementary entities* are identified and grouped into *elementary entity types*. An elementary entity is a ''thing'' that can be distinctly identified; there is no ambiguity concerning its existence or identity. An entity type is a set of entities sharing similar characteristics.

Then, in the second step, *relationships* among entities are identified, and also classified into *relationship types*. Relationships of the same type relate entities drawn from the same set of entity types.

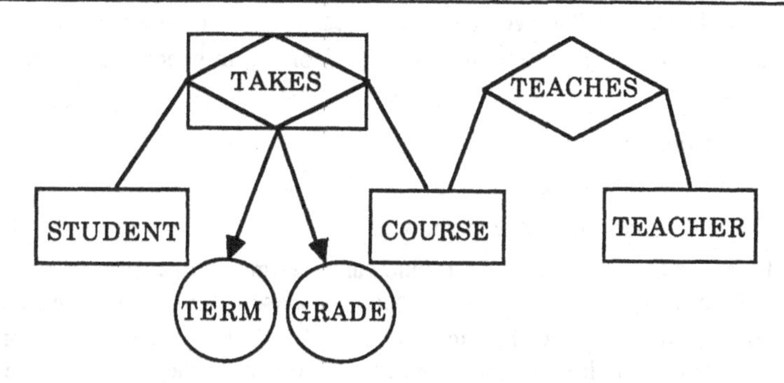

Fig. 2 An E-R diagram showing the schema of an academic record database.

A *composite entity* is an entity formed by other entities (elementary entities or composite entities); i.e., a composite entity cannot exist by itself whereas an elementary entity type can. A composite entity is similar to a relationship. The difference between them is that the former can have a property of its own, but the latter cannot.

A set of properties may be associated with each entity type (elementary or composite) to characterize entities of that type. Each property can be expressed in terms of *attribute-value pairs*. For example, in the statement "the AGE of a person x is 28," "AGE" is an *attribute* of x, and "28" is the *value* of the attribute AGE. As with entities and relationships, values are classified into *value types*.

Another way to look at an entity property is to consider it as a set-valued function from an entity type to the value type associated with the property. This aspect of attribute value assignment will be considered in more detail later.

The set of entity types and the set of relationship types together with associated properties constitute the conceptual schema of a database.

One advantage of the E-R method is the use of a diagrammatic notation, know as an *entity-relationship diagram* (E-R diagram). Fig. 2 shows an E-R diagram for the schema of a simple academic record database. We have three elementary entity types STUDENT, COURSE and TEACHER, one composite entity type TAKES, and one relationship type TEACHES. TERM and GRADE are properties of the composite entity type TAKES. Arrows from TAKES to TERM and GRADE indicates the functional nature of the two properties. For each instance of a STUDENT taking a COURSE, values for TERM and GRADE are functionally determined.

3.2. MODELING OF FUZZY DATA

3.2.1. Attribute Value Assignment. In the example of Fig. 2, it is assumed that exactly one attribute value is determined for each of TERM and GRADE. It is not always the case that the attribute values can be decided in a functional manner.

Suppose that an entity type E is characterized by a property P, and that the attribute and the value type associated with P are A_P and V_P, respectively. For each member e of type E, $A_P(e)$ returns its attribute value(s). The domain $Dom(A_P)$ of A_P, in general, consists of three types of values:[3]

Type 1: an atomic value;

Type 2: a non-atomic value; and

Type 3: a null value (i.e., value unknown).

The **type 1** implies the usual case of attribute value assignment as demonstrated in the example of Fig. 2. When designing a logical (relational) database schema, this type of attribute will be concatenated with the entity identifier to form a single relation.[4] The domain consisting entirely of **type 1** values will require no further consideration, and, in the following, we concentrate on the **types 2** and **3** of attribute value assignment.

The **type 2** value implies a set-valued attribute. A set-valued attribute is a special case of *fuzzy set*-valued attribute. We will defer the discussion on this matter until the section *3.2.2.* When the value is unknown, we must resort to the use of nulls (**type 3**). But note that the use of a special null value is the result of imposing atomicity on the attribute values. If we lift this restriction, there will be a better way of handling nulls. In fact, if we adopt the concept of set-valued attributes throughout the database, a null will be represented as an empty set, $\emptyset$. This arrangement for null values will require little special consideration.

3.2.2. Fuzzy Property. Now we go back to the **type 2** assignment and generalize it to a case of fuzzy property. As noted above, the **type 3** assignment may be regarded a special case, i.e., an empty set assignment, of the **type 2** assignment. Hence, the following argument applies to the **type 3** assingment as well as to the **type 2** assignment.

Suppose an entity e of type E exhibits a fuzzy characteristics with respect to a property P. Then the attribute value $A_P(e)$ is a possibilistic distribution over the value type V_P.

In previous approaches to modeling fuzzy property, they attempted to model a possibilistic distribution, as well as a null value, using a single construct, i.e., a relation. The relational model is a record-based model and performs well when applied to the data of a fixed format. As we observed earlier, however, the three types of attribute values differ diversely in their nature. Also, a possibilistic distribution is essentially a fuzzy set and has its own structure and carries semantics. But the relational model is rather poor in capturing semantics. These discussions conclude that, despite the current trend of developing a fuzzy relational database, the relational model is not a prime candidate for modeling a fuzzy property. The flaws of the fuzzy relational data model demonstrated in Section 2 may be attributed to this point.

A fuzzy set F is characterized by a membership function μ_F. In order to fully model a fuzzy set, its membership function must be made explicit through the modeling process. In the relation-based representation of a fuzzy set (Fig. 1), a property P with the value type V_P is considered only subordinate to an entity type E. While a whole relation is allocated to

---

[3]Here we assume that the property is applicable to all members of the entity type. If we are to allow for the case of "Property inapplicable", we will have to introduce a special type of null value to indicate inapplicability, and decide on the special treatment of such nulls. This issue by itself constitutes a separate research topic.

[4]The first normal form of the relational theory assumes the atomicity of attribute values. The case of **type 1** value assignment is equivalent to this normalization assumption in the relational theory.

represent E, only one column of that relation is given to the property P.

In our approach, we treat the entity type E and the value type V_P equally so as to make the membership function explicit (Fig. 3(a)). The value type V_P is considered a special kind of entity type, and the elements of E and the elements of V_P are linked to each other according to the attribute assignment by A_p. For each e in E, we have a fuzzy set-value assignment $A_p(e)$, and its membership function is given by

$$\mu_{A_p(e)}: V_P \rightarrow [0, 1].$$

In the figure, the membership grade is indicated by a weight assigned to each instance of a relationship between E's elements and V_P's elements. So, in Fig. 3(a), we have:

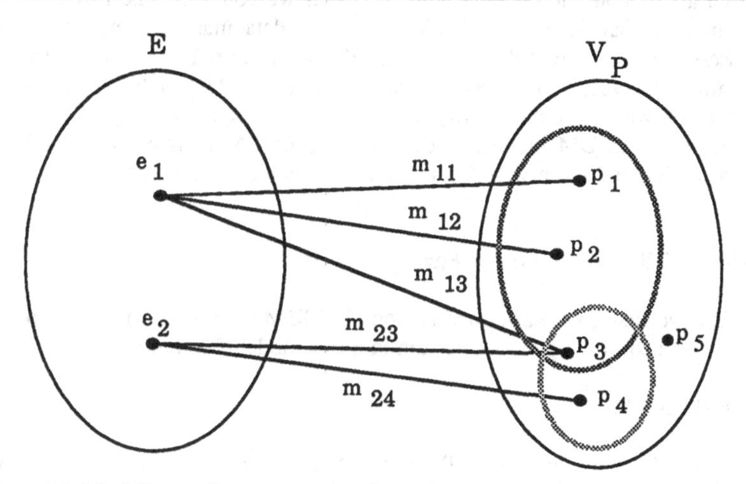

(a) Modeling a fuzzy set-valued attribute using the E-R model.

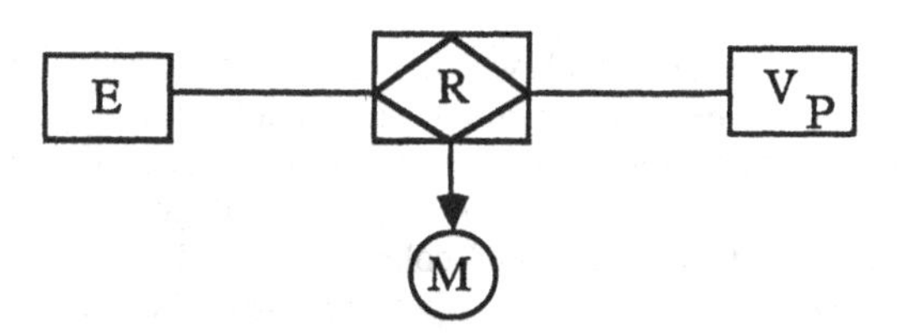

(b) An E-R diagram for (a).

Fig. 3 Modeling a fuzzy property using the E-R model.

$$A_P(e_1) = \{ m_{11}/p_1, m_{12}/p_2, m_{13}/p_3 \}, \text{ and}$$
$$A_P(e_2) = \{ m_{23}/p_3, m_{24}/p_4 \}.$$

If we model the situation depicted in Fig. 3(a) in terms of entity and relationship, we obtain an E-R diagram as shown in Fig. 3(b). We have two elementary entity types E and V_P and a composite entity type R with an attribute M. In this E-R schema, the membership function manifests itself as a composite entity type R with the attribute M.

To sum up, we have raised the status of a fuzzy property P from a mere attribute of an entity type E to a separate independent entity type V_P. By doing so, the membership function $\mu_{A_P(e)}$ associated with the fuzzy property P has come to be explicitly represented as a composite entity type relating the two entity types E and V_P.

Now we are in the position to transform the E-R diagram of Fig. 3(b) into a logical database schema of an appropriate model. It would be best if we could use the E-R model throughout the design process, but there is no widely accepted data manipulation facility for the E-R model. Observation of the E-R diagram reveals that only the binary relationships are employed to model a fuzzy property. The class of the E-R model in which only the binary relationships are allowed is known as a binary entity-relationship model (BERM) [4]. So, in the next section, we will turn to EGDM for further development; EGDM is as powerful as BERM in data definition capability, and has a well-defined rich data manipulation facility.

4. EGDM-based Representation of Fuzzy Property

In this section, we first give a brief description of EGDM (Section 4.1), and then transform the E-R schema obtained in the previous section into this model (Section 4.2).

4.1. GDM AND EGDM

Extended Graph Data Model (EGDM) [8, 9] is the result of enhancement on Graph Data Model (GDM) proposed in 1983 by Kunii [15]. It is implemented using a front-end data model processor [13] on top of a GDM-based database management system called G-BASE [18].

GDM is a link-oriented data model which belongs to a family of *post-relational* data models. It has been applied to some advanced application areas such as knowledge handling [16] and user interface construction [19]. Structural elements of GDM (and also of EGDM) consist of *record types* and *link types*. A record type is analogous to a relational scheme in the relational model, and a link type is a named collection of bidirectional indices between a pair of record types. In GDM, only the record types are allowed to have attributes of their own. But EGDM allows attributes on both a record type and a link type.

The most marked difference between the relational model and EGDM/GDM is in the introduction of *link types* as a structural element of the model. In the relational model, all that is visible to the users is the data organized in the tabular form; hiding everything else contributes to the model's simplicity and flexibility, but, at the same time, limits the descriptive power of the model and the performance of relational systems.

GDM introduced link types for the purpose of making access paths explicit to the users to

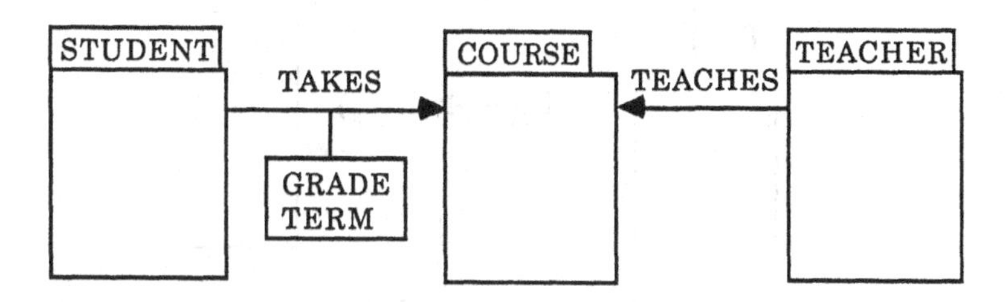

Fig. 4 The schema graph equivalent to the E-R diagram of Fig. 2.

obtain higher performance. Later it was discovered that link types also contribute to the modeling capability of the model, and this observation led to the development of EGDM by allowing link attributes.

An EGDM database schema can be illustrated using a *schema graph.* Fig. 4 shows a schema graph for the academic database used in the example of Fig. 2. A rectangle corresponds to a record type with its name indicated in the tag attached to each rectangle. If there are any attributes associated with an entity type, they may be listed inside the rectangle. An arrow connecting record types represents a link type; the direction of an arrow indicates the direction in which each link type is defined.[5] A rectangle attached to a link type indicates *link attributes*, if any. In a schema graph, each record type corresponds to a node in a di-graph, and each link type to an arc. The attributes attached to a link type then corresponds to a weight assigned to an arc of the di-graph.

Comparison of the two figures, Figs. 2 and 4, suggests a close correspondence between an E-R diagram and a schema graph of EGDM. An E-R diagram incorporating only binary relationship types can be directly transformed into a schema graph. For each elementary entity set we have a record type, and for each relationship type we have a link type. In case of a composite entity type, we have a link type with attribute(s).

4.2. TRANSFORMING THE E-R SCHEMA

Now we describe the situation modeled in Fig. 3 using a schema graph. The E-R diagram consists of two elementary entity types and one composite entity type. So the equivalent schema graph should contain two record types and one attributed link type defined between them (Fig. 5).

An entity e of type E is modeled as a record belonging to a record type R_E corresponding to E. The value type V_P is also transformed into a record type R_{V_P}. Then, the relationship

[5]Since a link type can be traversed in either direction, the directedness of a link type does not in any way limit the data manipulation capability of EGDM. In fact, it helps make the semantics of each link type clear (e.g., "A STUDENT TAKES a COURSE" rather than "A COURSE TAKES a STUDENT"). So the directed link type should be considered an asset to the model rather than a liability.

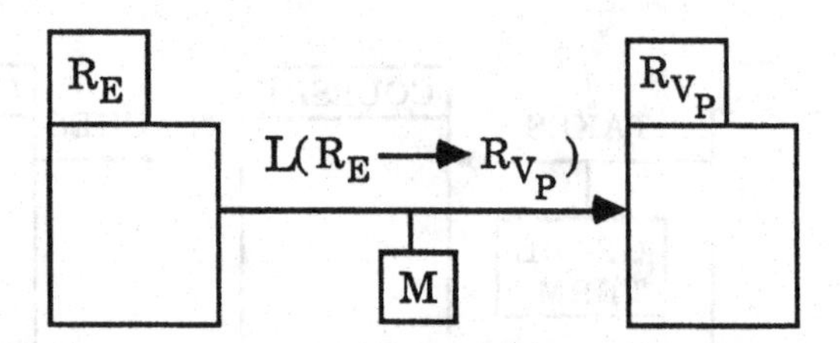

Fig. 5 A schema graph for modeling a fuzzy set-valued attribute.

between E and V_p naturally leads to a link type L from R_E to R_{V_p}. An attribute M is assigned to L for storing the values of the membership function $\mu_{A_p(e)}$.

5. Examples

As the first example of our graph-based formulation, let us consider the database relation of Fig. 1 again. The situation depicted here can be modeled in our approach as shown in Fig. 6. By separating the attribute "# of children" as a separate entity, both unnormalization and null values are avoided.

Next, we demonstrate the data manipulation facility of EGDM as applied to fuzzy data using

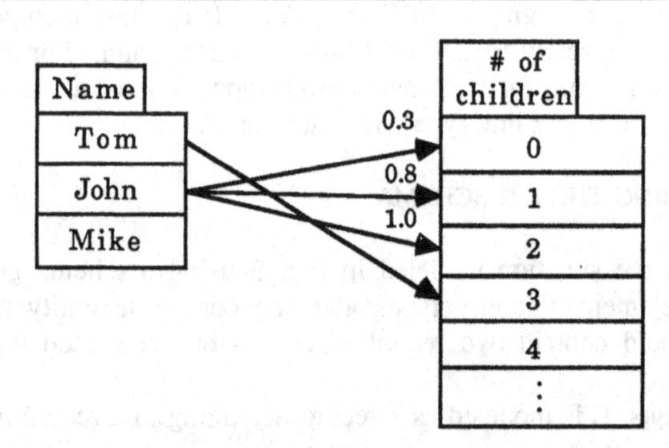

Fig. 6 EGDM-based representation of a fuzzy property.

a slightly more complex example. Let us consider a database storing the information on people's ages. To construct this database, two record types PERSON and AGE are required. AGE represents a property of the entity type PERSON. We establish a relationship between PERSON and AGE in the form of a link type YEARS, which accounts for possibilistic distributions concerning people's ages.

In addition, we introduce a third record type AGE_GROUP to account for natural language expressions (linguistic variables) on ages, such as "young" and "old". Since AGE_GROUP may be viewed as another property of PERSON, a link type LOOKS is defined between AGE_GROUP and PERSON. Note that LOOKS represents an atomic characteristic of PERSON, that is, an element of PERSON exhibits an atomic characteristic with respect to AGE_GROUP. The link type BREAKDOWN between AGE_GROUP and AGE models

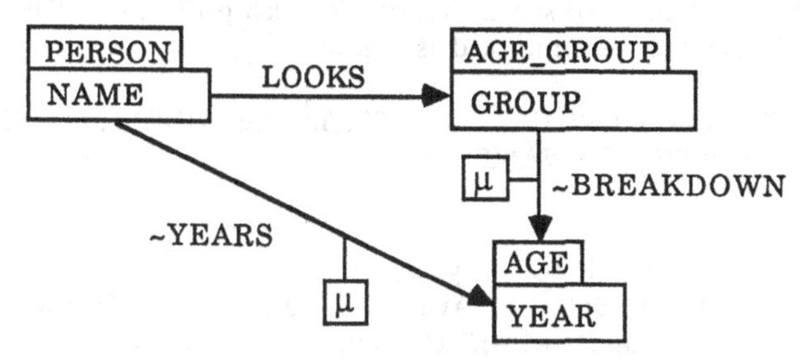

(a) The schema graph for the age database.

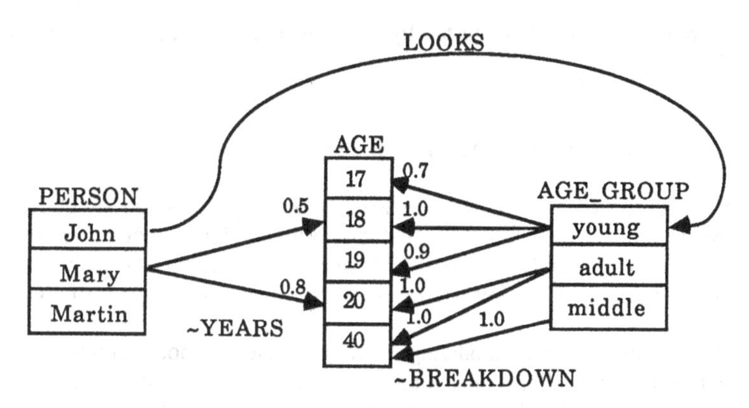

(b) An instance of the age database.

Fig. 7 An example: the age database.

possibilistic distributions expressed by linguistic variables. When processing a query, each instance of a (non-fuzzy) atomic characteristic can be considered a fuzzy subset with just one element at the full membership grade (i.e., 1).

The schema graph of the database is shown in Fig. 7(a). Inside each rectangle are shown attributes of each record type. A link type which represents possibilistic distributions is signified with a tilde "~" as a prefix, and has an attribute μ for storing membership grades.

Fig. 7(b) shows an instance of the age database. Each instance of record type and link type is shown with its attribute value. The absence of links starting from "Martin" indicates that his age is unknown; in the relation-based representation, a case like this will require null values.

Now let us consider a simple query on the instance shown in the figure: "Who is young with the grade of being young at 0.5 or more?" To answer this query, we first observe that there exist two access paths between PERSON and AGE_GROUP. Therefore, the above query must be broken down into two subqueries, one for each path. Using the syntax of the EGDM query language, they may be formulated as follows:

```
for AGE_GROUP[GROUP = "young"] ../inverse LOOKS/.. PERSON
        print PERSON.NAME;
>> John

for AGE_GROUP[GROUP = "young"]
        ../~BREAKDOWN/..  AGE ../inverse ~YEARS/.. PERSON
                [min(~BREAKDOWN.μ, ~YEARS.μ) >= 0.5]
        print PERSON.NAME;
>> Mary
```

The output from the system is indicated by the preceding ">>"; so the answer to the query is "John and Mary."

6. Conclusion and Future Research Directions

This paper has proposed a formulation of fuzzy sets using the graph structure, and realized it on a link-oriented data model EGDM. The new formulation does not suffer from the problems of unnormalization and null values which were inevitable in the previous relation-based approaches. The result of this research also demonstrates the richness of EGDM's modeling capability.

In this formulation, the same modeling construct is used to model entity types and their properties. Hence, the two different items are treated at the same level of abstraction. In reality, however, properties can only be meaningful in association with some entity type, which means that they are at different levels of abstraction. This issue calls for more detailed study on the nature of fuzzy data and the modeling techniques.

Once the structural organization suitable for fuzzy data is decided, there will be a variety of topics that open up for further investigation. Those topics include, for example, the definition of a data definition language and a data manipulation language and the development of a design theory for fuzzy databases.

Acknowledgement

Dr. H.S. Kunii, the director of Software Research Center of Ricoh Co., Ltd., kindly made their product G-BASE‡ available to the authors for the purpose of this research. Prof. S. Miyamoto of University of Tsukuba gave the authors motivation for research on fuzzy information processing.

Mr. Martin J. Dürst of Kunii Laboratory provided a valuable comment to improve the presentation style.

Reference

[1] Buckles, B.P., and F.E. Petry, "A Fuzzy Representation of Data for Relational Databases," *Fuzzy Sets and Systems*, Vol. 7, pp. 213-226, 1982.

[2] Chen, P.P., "The Entity-Relationship Model - Toward a Unified View of Data," *ACM Trans. on Database Systems*, Vol. 1, pp. 9-36, 1976.

[3] Chen, P.P., *Entity-Relationship Approach to Logical Data Base Design*, Monograph Series, QED Information Sciences, Inc., Wellesley, Massachusetts, 1977.

[4] Chen, P.P., "A Preliminary Framework for Entity-Relationship Models," in P.P. Chen (ed.), *Entity-Relationship Approach to Information Modeling and Analysis*, pp. 19-28, North-Holland, Amsterdam, 1983.

[5] Chen, P.P., "Database Design Based on Entity and Relationship," in S.B. Yao (ed.), *Principles of Database Design, Vol. I: Logical Organization*, Prentice-Hall, Englewood Cliffs, New Jersey, 1985.

[6] Codd, E.F., "A Relational Model of Data for Large Shared Data Banks," *Comm. ACM*, Vol. 13, pp. 377-387, 1970.

[7] Date, C.J., *An Introduction to Database Systems, Vol. II*, Addison-Wesley, Reading, Massachusetts, 1983.

[8] Fujishiro, I., Y. Shirai, H.S. Kunii, and T.L. Kunii, "Extending External Views on a Link-Oriented Data Model," *Proc. 11th Annual Int. Computer Software & Applications Conf.*, pp. 397-402, IEEE Computer Society Press, 1987.

[9] Fujishiro, I., *A Logical Design Methodology for Link-Oriented Databases Using Data Model Mapping*, Doctoral dissertation, Department of Information Science, the University of Tokyo, 1988.

[10] Gaines, B.R., "Logical Foundations for Database Systems," *Int. J. Man-Machine Studies*, Vol. 11, pp. 481-500, 1979.

[11] Grant, J., "Null Values in a Relational Data Base," *Information Processing Letters*, Vol 6, pp. 156-157, 1977.

[12] Grant, J., "Partial Values in a Tabular Database Model," *Information Processing Letters*, Vol 9, pp. 97-99, 1979.

‡G-BASE is a trademark of Ricoh Co., Ltd. The product is currently available under the name "RICOHBASE."

[13] Ichikawa, Y., I. Fujishiro, Y. Shirai, H.S. Kunii, and T.L. Kunii, "Design and Implementation Issues of the Extended Graph Data Model –Development of an EGDM Double Precompiler–," *Proc. 36th Annual Convention IPS Japan*, 2F-5, pp. 493-494, Information Processing Society of Japan, 1988 (in Japanese).

[14] Kunii, T.L., "DATAPLAN: An Interface Generator for Database Semantics," *Information Sciences*, Vol. 10, pp. 279-298, 1976.

[15] Kunii, H.S., *Graph Data Language: A High Level Access-Path Oriented Language*, Ph.D. Dissertation, Department of Computer Science, the University of Texas at Austin, 1983.

[16] Kunii, H.S., "DBMS with Graph Data Model for Knowledge Handling," *Proc. 1987 Fall Joint Computer Conference*, pp. 138-142, IEEE Computer Society Press, 1987.

[17] Maier, D., *The Theory of Relational Databases*, Computer Science Press, Rockville, Maryland, 1983.

[18] Ricoh Co., Ltd., *G-BASE System Manual*, 1988 (in Japanese).

[19] Shirota, Y., Y. Shirai, and T.L. Kunii, "Sophisticated Form-Oriented Database Interface for Non-Programmers," in T.L. Kunii (ed.), *Visual Database Systems*, pp. 127-155, North-Holland, Amsterdam, 1989.

[20] Umano, M., "Implementation of Fuzzy Relational Algebra in Fuzzy Databases," *Technical Research Report*, Vol. 86, No. 192, DE-86-4, the Institute of Electronics and Communication Engineers, Japan, 1986 (in Japanese).

[21] Zemankova-Leech, M. and A. Kandel, *Fuzzy Relational Data Bases - A Key to Expert System*, Verlag TUV Rheinland GmbH, Köln, Germany, 1984.

[22] Zvieli, A. and P.P. Chen, "Entity-Relationship Modeling and Fuzzy Databases," *Proc. IEEE 2nd Int. Conf. on Data Engineering*, pp. 320-327, IEEE Computer Society Press, 1986.

Imprecision in Human Combination of Evidence

Marcus Spies
IBM Scientific Center
Tiergartenstr. 15
D - 6900 Heidelberg[1]

Abstract For inference problems involving combination of evidence with either fuzzy quantifiers or support intervals some basic properties are reviewed. An empirical study is reported that was set out to investigate consensus of human inference with model inferences. It is shown that consensus is heavily influenced by the factors positiveness, monotonicity, and entropy / nonspecificity. On the one hand, human heuristics distort consensus; on the other hand, specific model assumptions could be reformulated. An approach using conditional objects to this reformulation is outlined.

Introduction

A basic requirement for uncertainty management in expert and decision support systems is the presence of algorithms for combining evidence. However, it is open to question how such algorithms can consistently operate on partially consistent data provided by experts and to what degree users might legitimately feel the system's solutions to differ from their own expectations.

These problems gain in significance if one takes into account the results presented in the vast psychological literature on human "heuristics and biases" in processing any kind of uncertain evidence (Kahneman, Slovic, & Tversky (eds.), 1982; for a recent review, see Rachlin, 1989, ch. 2-4; for an explanatory approach in terms of cognitive strategies, see Scholz, 1986).

One might be tempted to assume that, extending uncertainty modelling beyond the limits of the classical probability calculus, one might find a more natural explanation of such heuristics and biases in terms of overly delimited model assumptions on the part of classical probability taken as descriptive standard of reference. The present investigation was set out to test precisely this assumption. Two models that generalize probabilistic calculus were selected: Zadeh's inference mechanism involving fuzzy quantifiers (Zadeh, 1985) and Baldwin's support logic (Baldwin, 1986). In the sequel, this paper assumes familiarity with the essential notions and formalisms underlying each model of uncertainty.

---

[1] On leave from Institute for Psychology; Free University of Berlin; Habelschwerdter Allee 45; D - 1000 Berlin 33

W. H. Janko et al. (eds.), Progress in Fuzzy Sets and Systems, 161–175.

The basic idea of an empirical investigation of the assumption was to take some meaningful properties of inference problems involving uncertainty and to measure whether they act significantly on the degree of consensus between human inferences and model solutions. The domain of inference problems contained the four classes of syllogisms defined in Zadeh's paper (1985); in the present paper, however, I wish to focus on problems involving combination of evidence only (for a full acount of the experiments and more theoretical background, see Spies, 1989, in press).

Human combination of evidence implicitly takes place in many intellectual activities; expert knowledge and everyday cognitive tasks both make use of it. Yet it seems hard to capture the essence of these activities in mathematical models. The reason is that the limitations of most of the existing models, namely the assumption of compositionality (see Zadeh, 1985) and different independence assumptions, are hard to verify as for their relevance to human cognition. By contrast to combination of evidence, problems like chaining only involve assumptions concerning what you know in a *single* premise, but no assumptions concerning the relationship of the information given in *several* premises. Therefore, in order to come to grips with human combination of evidence, an empirical approach seems necessary in the first line. Rather than attempting to build a(nother) prescriptive model on combination of evidence, the aim of the present research is to compare human combination of evidence as to its consensus with outputs generated from existing models. More specifically, it should prove helpful to knowledge engineers and systems designers to establish whether some specific properties of uncertainty model expressions can be demonstrated to increase this consensus.

It has been shown that the different kinds of uncertainty modelled in these two approaches cannot be reduced to one (Spies, 1987). Therefore, the models are not directly compared to each other in this investigation.

Two ***modi of combination of evidence*** were empirically investigated. In Zadeh's terms (Zadeh, 1985), these are the antecedent and the consequent conjunction syllogisms, respectively. In order to introduce the underlying inference problems in a succint manner I propose the following notation: Read R(B/A) either as fuzzy proportion of B's among A's, A and B being (crisp or fuzzy) subsets of a universe U, or as support pair of the rule "if B then A", A and B being propositions in a Boolean algebra; in both cases, let AC denote conjunction of A and C. In the *antecedent conjunction problem* you are given R(B/A) and R(B/C) and you have to conclude R(B/AC). This is the classical problem of a diagnostic query: Given two "symptoms" (or something analogous) pointing each to the same disease to different degrees of quantitative or logical evidence, what is the their joint effect in supporting the hypothesis of that disease?

In the *consequent conjunction problem*, using the same notation, you are given R(A/B) and R(C/B) and you are asked to evaluate R(AC/B). This problem arises when we have to combine categories to get information about a certain phenomenon.

Properties of Combination of Evidence in the two Models

In this section some properties of combining of antecedents with support intervals and fuzzy quantifiers will be proved that will turn out to have significant impact on consensus of human information processing with these formal models. Only properties of the antecedent conjunction problems are shown. As for the consequent conjunction syllogisms, the reader is referred to Spies (1989, in press).

Concerning fuzzy quantifiers, a simple result with respect to ordinal properties implied by the likelihood combination rule will be proved. Then, the behaviour of a fuzzy result of antecedent conjunction for strictly conflicting evidence will be studied.

As for support intervals, it will be shown that, using the Dempster combination rule, one ends up with a narrower interval than any interval given in the premises of the antecedent conjunction problem. However, the width of the resulting interval increases with the weight of conflict in the inference problem.

Fuzzy quantifiers will be written as quadruples $[a, b, c, d]$, where it is assumed that

1. $a \leq b \leq c \leq d$,
2. the membership function Q takes the following values: $Q(a) = 0$, $Q(b) = 1$, $Q(c) = 1$, $Q(d) = 0$.

Thus, if we assume the membership function to be linear the quadruple $[a, b, c, d]$ describes a *trapezoidal* fuzzy quantifier. It can easily be seen that the extension principle (see Dubois & Prade, 1980) implies that for all arithmetic operations it suffices to apply them to the cornerpoints of the operands in order to get the correct cornerpoints of the results. However, linearity of the flanks is not generally preserved by extended arithmetic opertations. The membership in the core of the resulting quantifier will be constantly one, whatever the operation was. Therefore, the quadruple notation gives us a good orientation in some intricate fuzzy arithmetic will be used.

According to the likelihood combination rule, the combined probability of (say, disease) D, given (say, symptoms) A and B, reads:

$$p(D \mid A \cap B) = \left[1 + \frac{p(\overline{D} \mid A)}{p(D \mid A)} \; \frac{p(\overline{D} \mid B)}{p(D \mid B)} \; \frac{p(D)}{p(\overline{D})} \right]^{-1}.$$

Note that this formula assumes conditional independence of the evidences A and B given the hypothesis or given the negated hypothesis (for more discussion concerning this formula, see Goodman & Nguyen, 1985, pp. 280 sqq.; Heckerman, 1986). For the following development, note that

$$p(X \mid Y) \geq p(Y) \;\Rightarrow\; p(\overline{X} \mid Y) \leq p(\overline{Y}).$$

Let us now assume that

$$p(D \mid A) \geq p(D),$$

which is equivalent to assuming that evidence A supports hypothesis D. The combination formula can be rewritten as

$$p(D \mid A \cap B) = \left[1 + \frac{p(\overline{D} \mid A)}{p(D \mid A)} \; \frac{\dfrac{p(\overline{D} \mid B)}{p(\overline{D})}}{\dfrac{p(D \mid B)}{p(D)}} \right]^{-1}.$$

Here, the numerator of the right-hand fraction is less than or equal to unity, by assumption; as a consequence, the denominator is at least equal to unity (recall the inequality relationship mentioned above). Hence, the whole expression in square brackets will not decrease if we replace the right-hand fraction by unity. Thus, we can write:

$$p(D \mid A \cap B) \geq \left[1 + \frac{p(\overline{D} \mid A)}{p(D \mid A)} \right]^{-1} = p(D \mid A),$$

where the last equality results from the definition of odds from probabilities. In a similar way, assuming

$$p(D \mid A) \geq p(A),$$

we conclude that $p(D \mid A \cap B) \geq p(D \mid B)$. Thus, we have proved the ***enhancement property*** of the likelihood combination rule: Assuming that both evidences support the hypothesis, we can conclude that their conjunction gives a stronger support than each support contributed by a single evidence. Formally,

$$\min(p(D \mid A), p(D \mid B)) \geq p(D) \;\Rightarrow\; p(D \mid A \cap B) \geq \max(p(D \mid A), p(D \mid B)).$$

By symmetry, it follows that if both evidences weaken the hypothesis, their combined support will be even weaker than each single contribution.

Next, let us discuss the consequences of combining strictly contradictory evidence with fuzzy quantifiers. Strictly contradictory evidence is given by fuzzy quantifiers like $Q(D \mid A) = [0, 0, a, b]$ and $Q(D \mid B) = [c, d, 1, 1]$. The crisp likelihood combination formula is not defined in a similar case. The *fuzzy likelihood combination rule* looks like this (Zadeh, 1985):

$$Q(D \mid A \cap B) = \left[1 \oplus Q(\overline{D} \mid A) \otimes Q^{-1}(D \mid A) \otimes Q(\overline{D} \mid B) \otimes Q^{-1}(D \mid B) \otimes Q(D) \otimes Q^{-1}(\overline{D})\right]^{-1}.$$

Using the usual extension of arithmetic on real numbers to $\mathbb{R} \cup \{-\infty, +\infty\}$, and Dubois' and Prade's (1980) lemmata concerning fuzzy arithmetic, we can evaluate the fuzzy likelihood combination rule for strictly contradictory evidence using the quadruple notation. For negations, we obtain:

$Q(\overline{D} \mid A) := [1-b, 1-a, 1, 1]$; $Q(\overline{D} \mid B) := [0, 0, 1-d, 1-c]$.

For reciprocal values of the quantifiers, we find:

$Q^{-1}(D \mid A) := [\frac{1}{b}, \frac{1}{a}, \infty, \infty]$; $Q^{-1}(D \mid B) := [1, 1, \frac{1}{d}, \frac{1}{c}]$.

Thus, the fuzzy prior odds of $\overline{D}$ given B and A, respectively, labelled X and Y for simpler reference, read as follows:

$$Q(\overline{D} \mid A) \otimes Q^{-1}(D \mid A) := [\frac{1-b}{b}, \frac{1-a}{a}, \infty, \infty] = X$$

$$Q(\overline{D} \mid B) \otimes Q^{-1}(D \mid B) := [0, 0, \frac{1-d}{d}, \frac{1-c}{c}] = Y.$$

Using again extended fuzzy arithmetic, it follows that

$X \otimes Y = [0, 0, \infty, \infty]$.

Comparing with the likelihood combination formula for the crisp case, it becomes obvious that any base rate Q(D) that is not monotonic (i.e. $Q(D) = [e, f, g, h]$ with $0 < e, 1 > h$) will not affect the result of the computation of the odds product in the fuzzified version with strictly conflicting evidence; i.e., in this case, we can write $X \otimes Y \otimes Q(D) \otimes Q^{-1}(\overline{D}) = X \otimes Y$. Completing computations, we can write:

$Q(D \mid A \cap B) := [1 \oplus X \otimes Y]^{-1}$

where

$1 \oplus X \otimes Y = [1, 1, \infty, \infty]$

and thus

$Q(D \mid A \cap B) = [0, 0, 1, 1]$.

We have obtained a quantifier ranging from 0 to 100% and giving all these percentages a constant possibility of one. By analogy to the Dempster/Shafer terminology, I would like to call it the vacuous quantifier. Thus, ***strictly conflicting evidence with fuzzy quantifiers leads to a vacuous quantifier.*** This is, it seems to me, an important and intuitively quite appealing result.

Let us now turn to properties of combination of evidence with Baldwin's support logic.

It will be proved that, using Dempster's rule, a shrinkage of support intervals from the sources to the combined belief function always occurs. Thus, combining evidence in an antecedent conjunction problem always results in narrowing down the width of the support intervals, thus getting closer and closer to probabilistic situation, where the support interval vanishes. It should be recalled that, since support pairs are derived from a dichotomous support function, the width of a support interval equals the mass committed to the whole frame of discernment (see Shafer, 1976). In terms of logic, this means that the width of the support interval equals the support committed to the tautology.

First, I show that, the higher the conflict, the greater the width of the combined support interval. Then, I show that the width of the resulting support pair is at most the minimum of the widths found in the two support intervals combined by Dempster's rule. Let, for i = 1 to 2, d_i denote the width $su_i - sl_i$ of the respective support interval $[sl_i, su_i]$. From the combination diagram (see Shafer, 1976; Baldwin, 1986; Spies, 1989, in press) it follows for the width of the combined support interval

$$d = Pl(p) - Bel(p) = \frac{d_1 d_2}{c}$$

where $c := 1 - sl1(1 - su2) - sl2(1 - su1)$. Now, Shafer (1976, p.65) defined the weight of conflict to be $\log 1/c$. Therefore, if the weight of conflict increases, leaving the widths of the premise support intervals unchanged, d will increase. This is the first result.

In order to derive the second result, one notes that c contains all of the probability mass committed to non-contradictory statements (or non-null sets) of the frame of discernment. Since d_1 equals the support of the first premise that is committed to the whole frame, there cannot occur any empty set in combining this support with any supports from the second premise. Therefore, it can be shown (Spies, 1989, in press, p. 53) that

$$c \geq d_1 + d_2 - d_1 d_2.$$

It follows, for $d_1 \neq 0$, $d_2 \neq 0$, that

$$d = \frac{d_1 d_2}{c} \leq \frac{d_1 d_2}{d_1 + d_2 - d_1 d_2} = \frac{1}{\frac{1}{d_1} + \frac{1}{d_2} - 1}$$

Now, if it were true that $d_1 < d$, one could deduce

$$d_1 < \frac{1}{\frac{1}{d_1} + \frac{1}{d_2} - 1} \Leftrightarrow 1 + \frac{d_1}{d_2} - d_1 < 1 \Leftrightarrow \frac{d_1}{d_2} < d_1$$

which is impossible, since $d_2 \leq 1$. The analogous contradiction follows from assuming $d_2 < d$. Thus, it follows that

$$d \leq \min(d_1, d_2),$$

which is the second result.

Taking together these two results, one can state that with Dempster's combination rule applied to support intervals, conflict raises "uncertainty" (where uncertainty is assumed to vary with the width of a support interval), but that this kind of uncertainty is always reduced from the premises to the result.

An empirical study on human versus model inferences

The experiment designed to shed more light on the consensus of human with model inference and thus to discuss human biases and heuristics in a new context will now be described.

It is claimed that in many applications people have to use numbers rather than words. Thus, instead of using linguistic variables, graphical representations of fuzzy quantifiers as possibility distributions of percentages and support pairs as voting patterns were given to the participants of the experiments. In a variety of problems, they were asked to edit the quantifier (support interval) corresponding to the conclusion of an inference problem. On the same screen were displayed the text and quantifiers (voting patterns) for the premises.

Three ***properties for factorial experimentation*** with fuzzy quantifiers and support pairs were defined.

Monotonicity A fuzzy quantifier is monotonic, if its possibility distribution takes the value 1 on the 0- or the 1-proportion (i.e., if the fuzzy quantifier can be understood as an imprecise "NONE" or "ALL"). Similarly, a support interval is monotonic if the necessary support is 0 or the possible support is 1 (i.e., when the underlying support function for a rule or a fact is simple instead of dichotomous).

Positiveness: A fuzzy quantifier is positive, if the core of its possibility distribution is a subset of the [0.5, 1] interval of proportions. Similarly, a support interval is positive if its necessary support is greater than the complement of the possible support (i.e., if the rule or fact is stronger supported than its negation).

Entropy/Nonspecificity: A fuzzy quantifier is imprecise if its entropy and nonspecificity lie beyond certain limits; in a similar way, imprecise support pairs are defined through the generalizations of entropy and nonspecificity measures for bodies of evidence (for an encompassing account of these measures and their properties, see Dubois & Prade, 1987).

> **Note:** This definition arbitrarily dichotomizes a continuous property of both fuzzy quantifiers and support intervals, if "certain limits" are chosen for a particular set of problems. Moreover, two distinct uncertainty measures are melted together. The reason for these simplifications is that experimental ressources were, as always, limited. A further study separately treating these different aspccts is currently being prepared.

The –scalar– ***dependent variable*** was defined such as to fulfill the following requirements:

- Continuity (i.e., insensitivity to small relative shifts in fuzzy quantifiers or support intervals).
- Robustness against overly delimited and overly diffuse data. A data fuzzy quantifier, for instance, is overly diffuse if the model fuzzy quantifier is wholly or to a major part contained in it, while the converse does not hold or only holds to a substantially lower degree.
- Maximum (unique) attained if and only if data fuzzy quantifier (support interval) identical to model fuzzy quantifier (support interval).
- Minimum (unique) attained if and only if data fuzzy quantifier (support interval) does not intersect with model fuzzy quantifier (support interval).

It has been shown (Spies, 1989, in press) that the following measure, named "degree of consensus" (DOC), fulfills these requirements. Let F_A denote a measure of interval or fuzzy interval A. Then, define "the part of A that belongs to B" as

$$F_{B|A} := F_{A \cap B} \div F_A$$

Then,

$$DOC(A, B) := F_{B|A} \times F_{A|B}$$

To the author's knowledge, no other existing measure of distance or possibility or similarity of (fuzzy) intervals fulfills all of the requirements previously listed. However, Zimmermann (1988, personal communication) pointed out that degrees of consensus are only applicable to fuzzy sets if they can be assumed to be of equal height.

Fuzzy quantifiers in the premises of experimental problems were always linear. In order to account for nonlinearity in the flanks of resulting quantifiers, two versions of

$$F_{B|A}$$

were calculated. The one assumed rescaled possibility axes and treated resulting quantifiers as linear, the other used an average from support length and core length of fuzzy quantifiers. (Calculation of the first version necessitated finding the basic feasible solutions of the set of linear inequalities describing the lines delimiting the convex intersection of two quantifiers.) Factorial effects were identical for both measures, indicating robustness of the effects reported.

For each (dichotomous) property P a three-level factor F(P) for syllogistic problems with two premises was defined by putting

F(P) = 0 if P = 0 in both premises;
F(P) = 1 if P = 1 in exactly one premise;
F(P) = 2 if P = 1 in both premises.

In terms of analysis of variance, the two components of any such factor F are obtained, respectively, by contrasting F(P) = 2 against F(P) = 0 and F(P) = 0 or 2 against F(P) = 1. In terms of syllogistic problems, the first component assesses the effect of presence or absence of the property described by P; the second assesses the effect of P's total presence or absence versus its partial presence. Note that, for positiveness, this last effect precisely corresponds to the effect of weakly versus strictly conflicting evidence. Therefore, the two effects of any factor F will be named "concordant" and "conflicting".

The ***experimental design*** for each modus within each model (fuzzy quantifier or support interval) then boiled down to a simple 3-factor factorial design with three levels of each factor. Repeated measurement was used to a wide extent to ensure acceptable cell sizes (for all details, see Spies, 1989, in press). Note that concordant and conflicting effects and their interactions have one numerator degree of freedom each; therefore, sphericity problems usually associated with covariance matrices from repeated measurement experiments are excluded.

Results - Discussion

Results for main effects are given in Table 1 on page 10. Interactions and raw deviations in corners of trapezoidal fuzzy quantifiers and interval limits of support intervals are discussed in Spies (1987).

- Human combination of evidence coincides considerably better with the results from the two models being used if uncertainty components of incoming information are monotonic and if they are "imprecise", i.e., if they exhibit a medium rather than low degree of nonspecificity and entropy. The effect of monotonicity is a clear hint for development of expert systems: All fuzzy quantifiers or support intervals that are not monotonic can be seen as intersections of monotonic ones. Obviously, it seems preferable to base expert and user interfaces on monotonic uncertainty components.

  The clear effect of the joint variation of nonspecificity and entropy makes empirical separation of the components of uncertainty (see Dubois & Prade, 1987) an urgent task for future research. A possible approach would consist of assessing utilities of imprecise data in a diagnostic framework. Thus, one would have to "scale" uncertain rather than risky options as in the von-Neumann-Morgenstern approach (see Jaffray, 1988; Spies, 1988).

- Conflicting evidence in support pair problems significantly deteriorates human performance compared with both the min-rule and the product-rule versions of Baldwin's (1986) theory. A similar problem does not occur with fuzzy quantifiers. It should be noted that in Baldwin's theory conflicting evidence is handled via renormalization with product rule and via reallocation under the minimum rule. Human performance falls somewhere in the middle between the entropy/nonspecitificty results according to both rules. In particular, the "narrowing-down"- property of Dempster's rule leads to the observed decreases of consensus in conflicting evidence.

		Antecedent Conjunction Syllogisms		**Consequent Conjunction Syllogisms**	
		Concordant	**Conflicting**	**Concordant**	**Conflicting**
Fuzzy Quantifiers	**Positiveness**	-- $F_{(1, 8)} = 78.16$ ***			-- $F_{(1, 6)} = 12.03$ *
	Monotonicity	+ $F_{(1, 8)} = 24.12$ **			
	Entropy/ Nonspecificity	+ $F_{(1, 8)} = 86.36$ ***	-- $F_{(1, 8)} = 25.37$ **	+ $F_{(1, 6)} = 25.29$ ***	
Support Pairs	**Positiveness**	-- $F_{(1, 14)} = 6.86$ *	-- $F_{(1, 14)} = 31.42$ ***	+ $F_{(1, 10)} = 13.06$ **	
	Monotonicity	+ $F_{(1, 14)} = 128.77$ ***	-- $F_{(1, 14)} = 26.44$ ***	+ $F_{(1, 10)} = 205.97$ ***	
	Entropy/ Nonspecificity	+ $F_{(1, 14)} = 379.19$ ***	-- $F_{(1, 14)} = 27.29$ ***	+ $F_{(1, 10)} = 73.32$ ***	

Table 1. Results from experiments on combination of evidence: Plus signs indicate improvements of consensus of human solutions with model solutions; minus sings indicate lower degrees of consensus if the corresponding property is present in at least one premise of the inferential problem; F-ratios for factorial effects are given together with numerator and denominator degrees of freedom; the number of stars indicates levels of significance.

- Common to both models are flaws in human processing of combining rather supportive ("positive") evidence. It may be assumed that humans combine evidences *without allowing for the base rate enhancement property*, or, more generally, without allowing for an archimedic property (i.e., that a large enough number of slightly positive rules with potentially different antecedents and identical consequent make up a near-to-full affirmation of the consequent). If that generalization could be corroborated by further empirical evidence, non-archimedic combination systems (see Hajek, 1985) could be appropriate if human behaviour were to be modelled.

- The data from the consequent conjunction problems reveal that human biases cannot be overcome in general by simply allowing for more uncertainty in the problem domain. The typical conjunction fallacy (see Kahneman, Slovic, Tversky (eds.), 1982) is mirrored in the positive effect of positiveness (concordant) for support intervals and in the deteriorating effect of conflicting positiveness for fuzzy quantifiers. Both effects can be traced back to an improper handling of conjunction by subjects.

To sum it up: We found reasons both for reformulating model assumptions and for remaining sceptical against the assumption that human biases are only artifacts produced by analyzing human cognition in terms of an overly restricitve model language. In the following section a possible way towards reformulating model assumptions will be outlined.

Conditional Objects and Human Combination of Evidence

If we try to understand the reasons for which consensus of human combination of evidence with results derived from formal models heavily depends on the factors used in the experiments, it should prove helpful to reexamine the very basic assumptions underlying the formulation of models. Now, a basic assumption common to reasoning with fuzzy quantifiers and support logic programming is the use of conditionals. Fuzzy quantifiers are fuzzy quantities defined on a conditional

$$\frac{B \cap A}{A}.$$

Similarly, Horn clause rules are conditionals of facts "on" other facts. Now, a very enlightening fact has been proved by Calabrese (see Nguyen, 1987): There is no set corresponding to a conditional. Precisely stated, there is no Boolean function on an algebra of subsets such that, for any probability measure P,

$$P(f(A, B, ...)) := \frac{P(A \cap B)}{P(B)}.$$

This fundamental insight means that it is not clear which set we are referring to when taking into account fuzzy quantifiers or support pairs in, for instance, combining evidence. One can view both fuzzy quantifiers and support pairs referring to rules as generalized conditional probabilities; however, the "generalization" only refers to the numerical part of the picture, but not to the set-theoretical part. As for the set-theoretical part, Nguyen (1987) has given some fundamental definitions. (A comprehensive discussion of this and alternative definitions can be found in Dubois & Prade (1988).)

More formally (see Nguyen, 1987), let $\mathscr{R}(+,\star)$ denote a Boolean ring of sets over a universe U; i.e., a commutative ring with unit element 1, null element 0, and such that

1. $a\star a = a \ \forall a \in \mathscr{R}$ (idempotency of every element)

2. $a + a = 0 \ \forall a \in \mathscr{R}$

Thus, + corresponds to disjoint union, while ★ corresponds to intersection.
The usual set union can, therefore, be written as $a \cup b := a + b + a\star b$ with the usual convention on operator priorities.
Complementation is denoted $\neg a := 1 + a$.

The conditional object $[b \mid a]$ is a set of elements x of $\mathscr{R}$ such that the intersection of x with a is the same as the intersection of b with a (Nguyen, 1987):

$[b \mid a] := \{x \in \mathscr{R} \mid x\star a = b\star a\}$.

Equivalently,

$[b \mid a] := \{x \in \mathscr{R} \mid x = a\star b + y\star\neg a \ , \forall y \in \mathscr{R}\}$.

$\mathscr{R}(+,\star)$ is a lattice. We denote the infimum and supremum of $[b \mid a]$ by $\sqcap [b \mid a]$ and $\sqcup [b \mid a]$, respectively.

Since $a\star b$ and $y\star\neg a$ can never intersect, it follows from the second definition formula that

1. $\sqcap [b \mid a] = a\star b$

2. $\sqcup [b \mid a] = a\star b + 1 + a = \neg a \cup b$, i.e. standard implication.

From the monotonicity of probability measures, it follows that

$p(x) \in [p(a\star b) \ , p(\neg a \cup b)] \ \forall x \in [b \mid a]$.

That means, a conditional object $[b \mid a]$ encompasses all sets implied by the conjunction $a \bigcap b$. It is not hard to see that these sets form a lattice. So, actually, what we obtain when being given a (fuzzy) relative quantity or a support pair associated with a rule is *evidence concerning a lattice of sets in a conditional object.*

Indeed, there is a very straightforward interpretation of the infimum and the supremum of this lattice that makes psychological consideration using conditional objects promising. The infimum of $[b \mid a]$ is the conjunction $a\star b$ itself, while the supremum of $[b \mid a]$ is the set $\neg a \bigcup b = 1 + a + ab$. If we are given any probability measure P on the boolean algebra of sets, we can say that $P(\sqcap [b \mid a])$ measures the degree to which we can expect to hit A's that are B's when randomly selecting individuals from U. Therefore, $P(\sqcap [b \mid a])$ measures the degree of verification of the statement "If X is A then X is B". On the other hand,

$1 - P(\sqcup [b \mid a]) = P(a \cap \bar{b}) = P(a + ab)$ measures the degree to which we will find X's that are A's but not B's, i.e., the *degree of falsification* of the statement "If X is A then X is B".

Conversely, $P(\sqcup [b \mid a])$ measures the degree to which the implication statement cannot be falsified. In a way, the degrees of verification and not-falsification can be viewed as counterparts of the degrees of necessary and possible support in Baldwin's support logic. The remaining probabilities of sets in $[b \mid a]$ measure degrees to which we hit sets that contain verifying and falsifying instances with respect to the implication statement. The closer to $P(\sqcap [b \mid a])$ these probabilities are, the more ***the evidence is concentrated on verification***. Conversely, the closer the probabilities lie around $P(\sqcup [b \mid a])$, the more ***the evidence is concentrated on failure of falsification.***

Let us now assume the following axioms for establishing a cognitive model on inference under uncertainty:

- The basic cognitive unit for processing a quantity is the set whose measure is given by this quantity.
- For any given information, there exists a cognitive tendency towards assuming a verification or failure-of-falsification mode of accepting or rejecting the information.

From these two axioms it can be concluded that evidence concerning a conditional object as a whole will be scattered over the elements of the conditional object (=sets) and that this distribution process will be controlled by a – until now undetermined – cognitive tendency towards verification or failure-of-falsification mode. Thus, *the result of a cognitive process of taking into account evidence concerning a conditional object, according to this model, is a valued lattice.*

It turns out that this model can explain some of the phenomena found in the empirical studies reported in the previous paragraphs with quite few additional assumptions.

In an antecedent conjunction problem, it can be shown that

$$\sqcap [b \mid a \star c] \leq \sqcap (\sqcap [b \mid a], \sqcap [b \mid c])$$

and

$$\sqcup [b \mid a \star c] \geq \sqcup (\sqcup [b \mid a], \sqcup [b \mid c]).$$

Thus, the span of the conditional object in combining evidence becomes larger. This also holds in the case of conflicting evidence. *Therefore, using conditional objects, we no longer have to stick with either the enhancement property or with the "narrowing-down"-property! Both properties can, but need not, hold with conditional objects.*

It can be shown (see Spies, 1989, in prep.) that the antecedent conjunction problem, if formulated numerically with conditional objects, has the analogous property. Thus, the interval in [0, 1] that corresponds to the span of the conditional object will become broader in the process of accumulating evidence. However, the concentration of evidence cannot be described in similar generality. With the combination method sketched in Spies (l.c.) compatibility with Dempster and Bayesian combination rules is secured. Subjects' data from the empirical studies actually lie almost certainly in such intervals corresponding specific combinations of conditional objects. Hence, the model is powerful enough to account for so-called human biases in combining evidence. This at least indicates the usefulness of a cognitive model using conditional objects.

The lack of conservatism for negative premises and the corresponding conservatism for positive premises in antecedent conjunction problems can, in turn, be explained if we assume the aforementioned cognitive tendency to operate according to the following rules:

- *If evidence is negative, it concentrates rather around lack of falsification.*
- *If, conversely, evidence is positive, it rather concentrates around verification.*

Assuming this rule to govern human processing of evidences in combining them, the distribution of evidence over the conditional object will produce precisely the phenomena found in the empirical study w.r.t. the enhancement property.

To summarize, the theory of cognitve modelling human inference under uncertainty with conditional objects seems to allow for convincing explanations of so-called human "biases" in combination of evidence. A full development of this theory will be given in a forthcoming paper (Spies, 1989, in prep.).

References

1. Baldwin, J. F. (1986): Support logic programming. In: A. Jones, A. Kaufmann, H.-J. Zimmermann (eds.): Fuzzy Sets Theory and Applications, NATO ASI Series, Dordrecht, D. Reidel, pp. 133- 170.
2. Dubois, D., Prade, H. (1987): Properties of measures of information in evidence and possibility theories. Fuzzy Sets and Systems, 24, 2, pp. 161-182.
3. Dubois, D., Prade, H. (1988): Conditioning in Possibility and Evidence Theoires - A logical Viewpoint. In: B. Bouchon, L. Saitta, R. Yager (eds.): Uncertainty and Intelligent Systems (Proc. Second IPMU, Urbino 1988), pp. 401-408.
4. Edwards, W. (1982): Conservatism in human information processing. In: Kahneman, Slovic, Tversky (eds.): Judgment under uncertainty, New York, Cambridge University Press, pp. 359-369.
5. Goodman, I.R. & Nguyen, H.T. (1985): Uncertainty Models for Knowledge-based Systems; Amsterdam, North Holland.
6. Hajek, P. (1985): Combining functions for certainty degrees in consulting systems. Int. J. Man-Machine Studies, 22, pp. 59-76.

7. Heckerman, D. (1986): Probabilistic Interpretation for MYCIN's Certainty Factors. In: L. Kanal, J. Lemmer (eds.): Uncertainty in Artificial Intelligence. Amsterdam, North Holland.
8. Jaffray, J.-Y. (1988): Application of linear Utility Theory to Belief Functions. In: B. Bouchon, L. Saitta, R. Yager (eds.): Uncertainty and Intelligent Systems (Proc. Second IPMU, Urbino 1988), pp. 1-8.
9. Kahneman, D., Slovic, P., Tversky, A. (eds., 1982): Judgment under Uncertainty: Heuristics and Biases. New York, Cambridge University Press. Nguyen, H.T. (1987): On Representation and Combinability of Uncertainy. Second IFSA Congress Preprints, Gakushuin University, Tokyo; pp. 506-509.
10. Rachlin, H. (1989): Judgment, Decision, and Choice. New York, Freeman.
11. Scholz, R. W. (1986): Cognitive Strategies in stochastic Thinking. Dordrecht, D. Reidel.
12. Spies, M. (1988): Specificity and entropy as attributes in reduction of uncertainty—Towards a psychological model of preferences in uncertain evidences. ORBEL4 –
13. Spies, M. (1989, in press): Syllogistic inference under uncertainty—An empirical contribution to uncertainty modelling in knowledge-based systems with fuzzy quantifiers and support logic. Munich, Psychologie Verlags Union. Fourth congress on quantitative methods for decision making of the Belgian Operations Research Society, Brussels.
14. Spies, M. (1989, in prep.): A model of imprecise quantification that accounts for human "biases". Paper to be presented at the 3rd International Conference of IFSA, Seattle, Washington, August, 1989.
15. Zadeh, L.A. (1983): A computational approach to fuzzy quantifiers in natural languages. Comp. & Maths. with Applic. 9, pp. 149-184.
16. Zadeh, L.A. (1985): Syllogistic reasoning in fuzzy logic and its application to usuality and reasoning with dispositions, Institute of Cognitive Studies Report 34. Also in IEEE Transactions, Vol. SMC-15, No.6, pp. 754-763.

FUZZY SET THEORY APPLICATION FOR FEEDFORWARD CONTROL OF ILL-DEFINED PROCESSES

D.STIPANICEV
Faculty of Electrical Engineering, Machine Engineering and Naval Architecture, ***University of Split***
R.Boskovica bb,58000 SPLIT,Yugoslavia

J.BOZICEVIC
Faculty of Technology, ***University of Zagreb***
Pierottieva 6,41000 ZAGREB,Yugoslavia

ABSTRACT. This article describes the principles of fuzzy feedforward control,suitable for feedforward control of ill-defined processes for which it is uneconomic or even impossible to develop a deterministic or stochastic process model and to apply the conventional feedforward control theory. Special attention is given to the experiments with fuzzy feedforward controllers and experimental results are presented.

1. Introduction

One of the oldest principle in regulation is certainly that of invariant control,i.e. the achievement of total or partial independence of the controlled system under examination from disturbances acting upon it. Thousands of years ago the first devices for controlling windmills ,constructed by the Arabs contained the elements of control for compensating the external load torque [1]. From that time till the modern age invariance control and especially feedforward control has been intensively studied and applied in numerous processes, but always on the basis of their deterministic or stochastic models.

For a certain number of processes, because of their complexity, limited knowledge or stochastic environment it is too expensive,or sometimes even impossible to develop a proper and valid deterministic or stochastic model and to apply the results of known, conventional feedforward control theory.

The fuzzy set theory has proved to be a suitable approach to modelling, simulation and control of these ill-defined systems. Many papers dealing with the application of the fuzzy set theory in feedback control of ill-defined systems have been published.

We have studied the disturbance decoupling and invariant control tasks, and used the methods of fuzzy set theory for feedforward control. With the proposed methods the area of application of the fuzzy control is widened to the invariant , feedforward control of ill-defined processes and a new field of research is opened.

This paper review our research activities and presents new experimental results obtained with the fuzzy feedforward controller.

W. H. Janko et al. (eds.), Progress in Fuzzy Sets and Systems, 176–188.

2. Fuzzy models of ill-defined processes suitable for feedforward control

Several representations of fuzzy models of processes have been proposed as suitable for feedback control tasks. For the feedforward control another form of model has to be considered. Also in real situations process variables may have nominal operating values, or special control channels may be established to realize control. We have studied all these cases [2,3,4]. Only the case used in experiments (Ch.5) will be explained here.

Distinguishing the real space of manipulated input (U), disturbance input (D) and controlled output (Y) the fuzzy model of the first order process can be expressed in the form of the complex

$$M = < D , U , Y , S^* > \tag{1}$$

where S^* is a fuzzy relation defined by the fuzzy mapping of ordinary sets

$$S^* : D \times U \times Y \rightarrow Y \tag{2}$$

with membership function $S^*(d,u,w,y)$, $d \in D$, $u \in U$, $w,y \in Y$.

The fuzzy set y^*_{n+1} of the controlled output at a discrete time moment (n+1)T (T being the sampling interval) may be evaluated by the fuzzy relational equation

$$y^*_{n+1} = S^* ® u^*_n ® d^*_n ® y^*_n \tag{3}$$

where u^*_n , d^*_n and y^*_n are appropriate fuzzy sets of the manipulated input, disturbance input and controlled output in discrete time moment nT , and ® is a relational-relational composition operator.

If in the process description the principle of superposition can be applied then by considering process behavior near to the nominal operating values d_N , u_N and y_N, the fuzzy model (1) can be changed into a form

$$M = < D , U , Y , P^* , R^* > \tag{4}$$

where P^* and R^* are fuzzy relations with membership functions $P^*(\Delta d,\Delta y)$ and $R^*(\Delta u,\Delta y)$ describing the relationship between the disturbance input increment Δd and the controlled output increment Δy and the relationship between the manipulated input increment Δu and the controlled output increment Δy.

The fuzzy set of the controlled output increment Δy^*_{n+1} due only to one of corresponding inputs may be evaluated by fuzzy relational equations

$$\Delta y^*_{n+1} = P^* ® \Delta d^*_n \tag{5}$$
$$\Delta y^*_{n+1} = R^* ® \Delta u^*_n \tag{6}$$

Δd^*_n and Δu^*_n are appropriate fuzzy sets of disturbance and manipulated input increments.

Fuzzy relations S^* or P^* and R^* can be obtained by one of the fuzzy identification procedures. In experiments we have used Pedrycz identification algorithm [5].

Equation (3) or equations (5) and (6) are essential parts of the fuzzy feedforward control algorithm.

3. Fuzzy feedforward control algorithm

Two types of fuzzy feedforward control are distinguished. The first one is on-line fuzzy feedforward control. In each discrete time moment nT, the fuzzy set of control which will bring the process as close as possible to the total disturbance decoupling state, is calculated on-line using the fuzzy model of the process. In the second method the control procedure is based on the fuzzy feedforward controller relation calculated before (off-line).

Equation (3) or equations (5) and (6) are essential parts of on-line fuzzy feedforward control. Theoretically for the fuzzy system described with the equation (3) the total disturbance decoupling will be obtained if for each discrete time moment nT the fuzzy set of the controlled output y^*_n is equal to the fuzzy set of the controlled output in the previous time moment (n-1)T , y^*_{n-1} , for all fuzzy sets of disturbance input. This means that in each time moment manipulated input , u^*_n , must be calculated from the equation

$$y^*_n = S^* \circledR d^*_n \circledR y^*_n \circledR u^*_n \tag{7}$$

If in the description of the process behavior equations (5) and (6) can be used the starting equation is

$$P^* \circledR \Delta d^*_n = R^* \circledR \Delta u^*_n \tag{8}$$

because total disturbance decoupling will be obtained if the magnitude of the controlled output change Δy^*_{n+1} is the same for both changes of manipulated input Δu^*_n and disturbance input Δd^*_n.

The fuzzy feedforward control action is obtained by solving the equation (7) for u^*_n or solving the equation (8) for Δu^*_n .The problem is that such fuzzy relational equations have not a unique solution , or even worse there are a lot of cases when the exact solution doesn't exist. We have analyzed these cases and developed suitable algorithms for approximate, numerical solution of such equations [2,6,7].

4. Fuzzy feedforward controller

The fuzzy feedforward controller consists of three main parts: process interface , fuzzy interface and fuzzy feedforward control algorithm. Both process and fuzzy interface have input and output part as it is shown in Fig.1.

The process input interface consists of the measuring devices, but when the variables are scarcely measurable, the ability of the operator to observe and express linguistically the actual values of variables should be considered. At the fuzzy input interface these numeric or linguistic data are represented with fuzzy sets. The fuzzy output interface converts a fuzzy set of manipulated input or change of manipulated input into a single real value which represents the control action. 'Mean-value method' or 'mean-of-maxima' can be used as an adequate technique in this interpretation procedure [2,6]. Through the output process interface this control action is applied to the process.

The principles of the fuzzy control algorithm are described in Ch.3. It is important to emphasis that in the case of the equation (8) the sign of the control action has to be determinate too. The control action will be minus, if the trends of the controlled output increment for the disturbance input only, and the trends of the controlled output increment for the manipulated input only are equal for the same trends of input variables , and vice versa. If the trends of the controlled output increment are equal for the opposite trends of input variables, the control action will be plus.

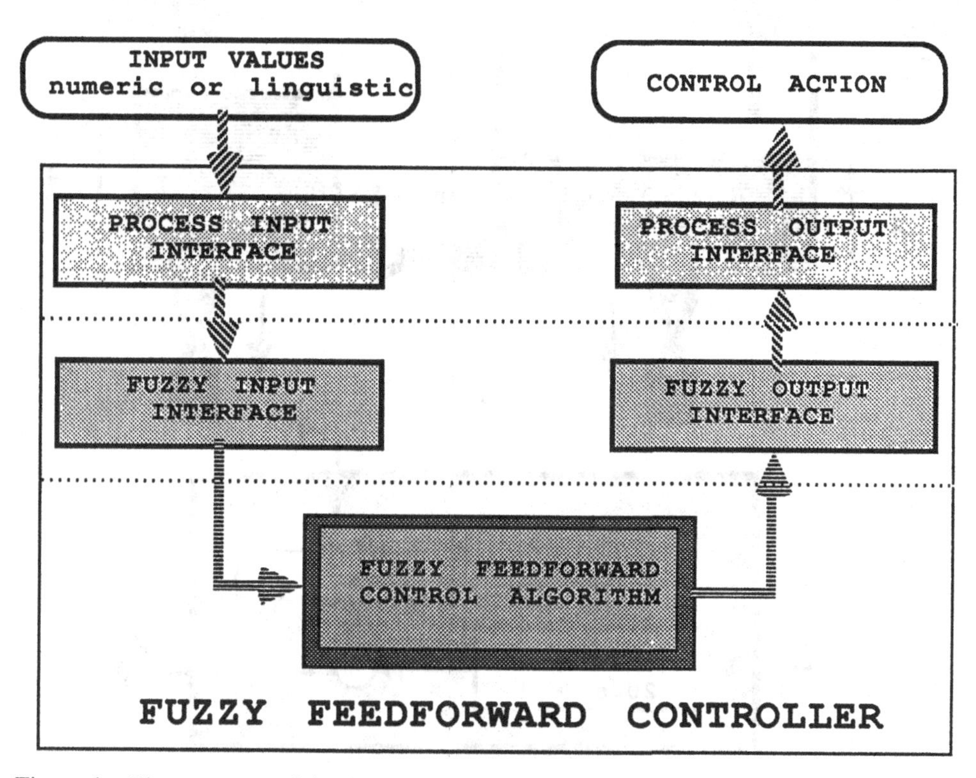

Figure 1. The structure of the fuzzy feedforward controller

After this short review of the fuzzy feedforward control principles, in the next section new experimental results obtained with fuzzy feedforward controller will be presented.

5. Experiments with fuzzy feedforward controller

In the first part of our experimental research, which will be described here, the process was simulated on the analog computer and the fuzzy feedforward control algorithm was programmed on the 8 bit on-line control microcomputer system. The considered controlled process was the coupled tanks water level process structured according to Fig.2., which is often used as a "benchmark" problem.The input flow to the Tank 1 was chosen as a manipulated variable , the level of the Tank 2 as a disturbance variable and the level of the Tank 1 as a controlled output variable. The deterministic representation of the considered process is shown on Fig.2. also.

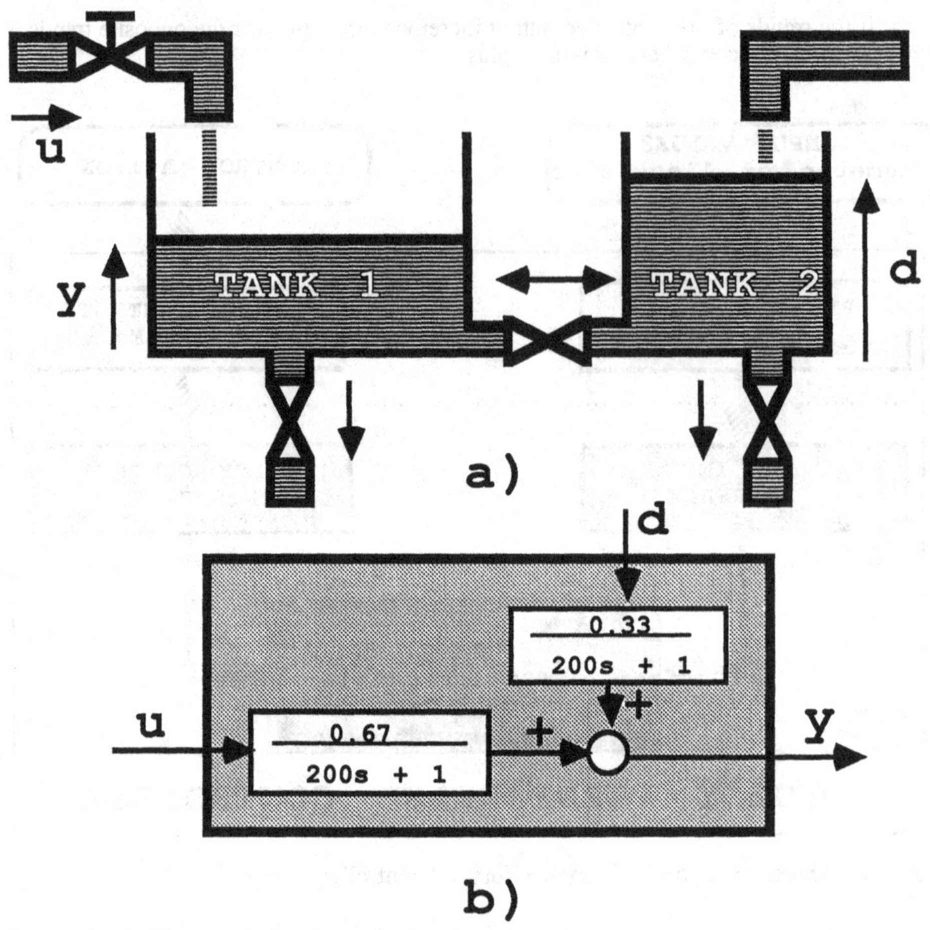

Figure 2. a) The coupled tanks water level process
b) The deterministic representation of the considered process

Generally ,the first stage in the process control is determination of the objectives of control. In this case the task is to maintain the water in the Tank 1 at the constant level and make them independent of the level in the Tank 2.

The second stage is determination of the process model. Normal operating values of the process variables in these experiments were $d_N = 5$, $u_N = 5$ and $y_N = 5$. The fuzzy model of the process was established using 14 subjectively observed, non-accurate, cause-effect numeric data:

$\{(\Delta d,\Delta y)\}=\{(-4.5,-0.18),(-2.1,-0.05),(-0.9,-0.015),(0,0),(1.3,0.05),(3,0.14),(5,0.21)\}$
$\{(\Delta u,\Delta y)\}=\{(-4.5,-0.42)\ ,\ (-2,-0.22)\ ,\ (-1,-0.15),\ (0,0)\ ,(0.5,0.1),\ (3,0.28),(4.7,0.43)\}$

The data were derived by human observation of variable values at the instruments where just zero, minimum and maximum were marked. Minimum (maximum) value of Δy was -0.5 (0.5) and of Δd and Δu -5 (5).

Each pair can be linguistically expressed with conditional statements, as for example the statement for pair $(\Delta d,\Delta y) = (-0.45,-0.18)$ is

"**If** the water level in the Tank 2 decreases 4.5 units below nominal value,
then the water level in the Tank 1 will decrease 0.18 units 30 seconds later"

The fuzzy model of the process must be in the form of two fuzzy relations $P^*(\Delta d,\Delta y)$ and $R^*(\Delta u,\Delta y)$. In order to unify the approach in such a way that it will be possible to use both numeric and linguistic data, a method of fuzzy discretisation of the real space is used [8]. It consists of cutting up the real space W into a series of non-disconnected fuzzy sets $Z^*_1, Z^*_2, \ldots, Z^*_I$, such that the entire space is covered. Real value $w_o \in W$ now can be represented with the fuzzy vector

$$\underline{Z}^*_{w0} = [\; Z^*_1(w_o) \;\; Z^*_2(w_o) \;\; \cdots \;\; Z^*_I(w_o) \;] \qquad (9)$$

and every fuzzy set A* which belongs to the space W and represents some linguistic data can be represented with the fuzzy vector

$$\underline{Z}^*_{A*} = [\; \sup_{w\in W}(\min(Z^*_1(w),A^*(w))) \;\; \ldots \;\; \sup_{w\in W}(\min(Z^*_I(w),A^*(w))) \;] \qquad (10)$$

Basic fuzzy sets $Z^*_1, \ldots, Z^*_I$ of fuzzy discretisation used in these experiments are shown on Fig.3. Each pair $(\Delta d,\Delta y)$ and $(\Delta u,\Delta y)$ is now transformed into pairs of fuzzy vectors $(\underline{\Delta d}^*,\underline{\Delta y}^*)$ and $(\underline{\Delta u}^*,\underline{\Delta y}^*)$. For example pair (-4.5,-0.18) is transformed into ([0.7 0.3 0 0 0 0 0], [0 0.78 0.92 0 0 0 0]). The linguistic labels of basic fuzzy sets are given on Fig.3. also.

A Pedrycz algorithm [5] was used as an identification algorithm suitable for calculation of fuzzy relations P^* and R^*, which are now fuzzy matrices $\underline{P}^*= [p_{lm}]$ and $\underline{R}^* = [r_{lj}]$. The algorithm is based on the clustering technique and minimisation of the sum of distances between the output of the model derived calculated with equations (5) and (6) and the collected fuzzy output data $\underline{\Delta y}^*$. The ISODATA algorithm [2] was applied as a clustering algorithm and the distance was specified as a Hamming distance. The minimal value of the performance index was obtained for seven clusters (each pair in its own cluster), and the final fuzzy matrices are

$$\underline{P}^* = [p_{lm}] = \begin{bmatrix} 0 & 0 & 0 & 0 & 0 & 0 & 0 \\ 0.08 & 0 & 0 & 0 & 0 & 0 & 0 \\ 1 & 1 & 0.09 & 0 & 0 & 0 & 0 \\ 0 & 0 & 0.69 & 1 & 0.16 & 0.16 & 0 \\ 0 & 0 & 0 & 0 & 0.3 & 1 & 0.80 \\ 0 & 0 & 0 & 0 & 0 & 0 & 0.2 \\ 0 & 0 & 0 & 0 & 0 & 0 & 0 \end{bmatrix} \qquad (11)$$

$$\underline{R}^* = [r_{lj}] = \begin{bmatrix} 0.52 & 0 & 0 & 0 & 0 & 0 & 0 \\ 0.49 & 1 & 0 & 0 & 0 & 0 & 0 \\ 0 & 0 & 0.68 & 0 & 0 & 0 & 0 \\ 0 & 0 & 0 & 0.4 & 0 & 0 & 0 \\ 0 & 0 & 0 & 0 & 1 & 0 & 0 \\ 0 & 0 & 0 & 0 & 0 & 0.68 & 0.42 \\ 0 & 0 & 0 & 0 & 0 & 0 & 0.58 \end{bmatrix} \qquad (12)$$

This fuzzy model can be easily converted into a linguistic model in which each rule has its degree of possibility. For example:

"Possibility that Δy is 'positive small' if Δd is 'positive big' is 0.8"
"Possibility that Δy is 'negative big' if Δu is 'negative big' is 0.52"

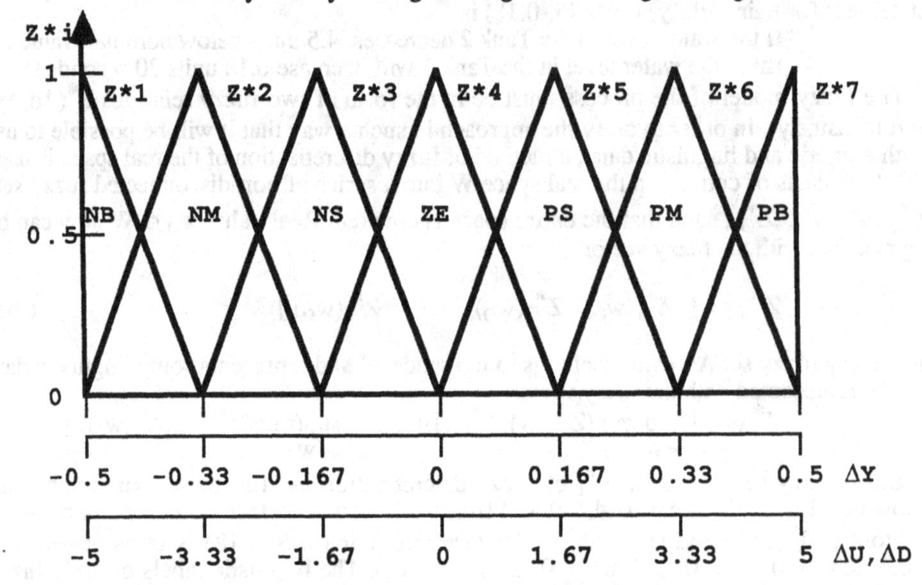

Figure 3. Basic fuzzy sets of fuzzy discretisation with their linguistic labels: NB-negative big , NM-negative medium , NS-negative small , ZE-zero , PS-positive small , PM-positive medium ,PB-positive big

The 'max-min composition' was used in the fuzzy feedforward control algorithm. The fuzzy control $\underline{\Delta u}^*$ was calculated either by α - composition as the appropriate inverse composition of max-min composition, or by numerical methods, depending whether or not the equation (6) has the exact solution [2,6]. Let us illustrate the calculation procedure with α - composition. For the fuzzy vector of disturbance input $\underline{\Delta d}^*=[d_m]$ the element u_j of the fuzzy vector of control $\underline{\Delta u}^* = [u_j]$ is obtained using the equation

$$u_j = \inf_l \{ \sup_m [\min (p_{lm},d_m)] \ \alpha \ r_{lj} \} \tag{13}$$

where for $a,b \in [0,1]$

$$a \ \alpha \ b = \begin{cases} 1 \ , \text{if } a \leq b \\ b \ , \text{if } a > b \end{cases} \tag{14}$$

In the fuzzy output interface the 'mean-value method' was applied to represent the fuzzy vector of control with unique value from the interval $\Delta U = [-5 , 5]$.

In order to compare the fuzzy approach with the conventional one, a linear regression model of the process is considered, too. For the same input-output data the regression model is

$$(\Delta y)_d = 0.0403 \cdot \Delta d \tag{16}$$
$$(\Delta y)_u = 0.0953 \cdot \Delta u \tag{17}$$

Fig.4. shows the process response for the conventional feedforward control derived from this regression model and fuzzy feedforward control for pulse, cosine and saw-tooth disturbance. The ideal situation (absolute or total invariance) is a horizontal line. Advantages of the fuzzy approach are evident. The sampling interval was 19.2 seconds.

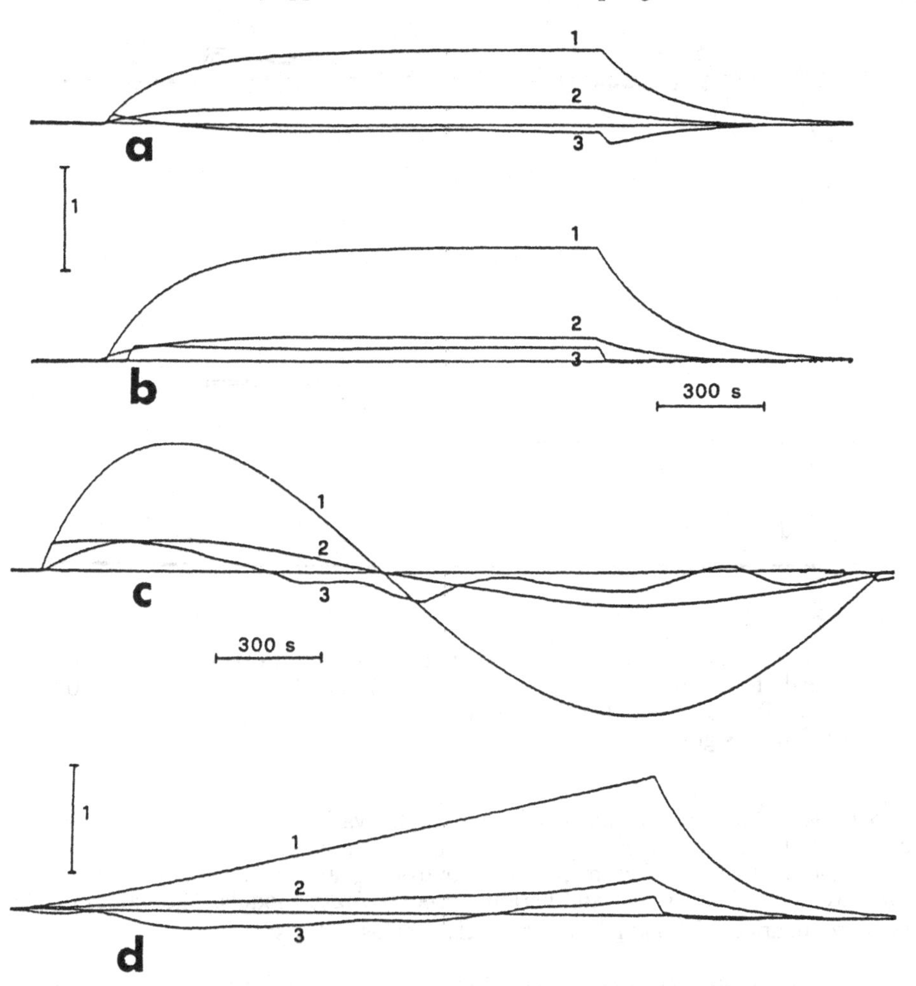

Figure 4. Process response without control (1), with conventional (2) and fuzzy (3) feedforward control for:
(a) Pulse disturbance 1400 seconds long with 2.5 units amplitude ,
(b) Pulse disturbance 1400 seconds long with 4 units amplitude,
(c) Cosine disturbance $\Delta d = 5 \cdot \cos 0.002 \cdot t$, and
(d) Saw-tooth disturbance $\Delta d = 0.0028 \cdot t$, 1800 seconds long.

Fig.5. shows the process responses for the fuzzy feedforward control , saw-tooth disturbance and different sampling intervals. The control signal is shown, too. The change of the sampling interval to approximately 300 seconds has no great influence on the process response, but even for T = 560 seconds the response is still satisfactory, although the magnitude of the control signal has been changed during the whole disturbance occurrence only four times .

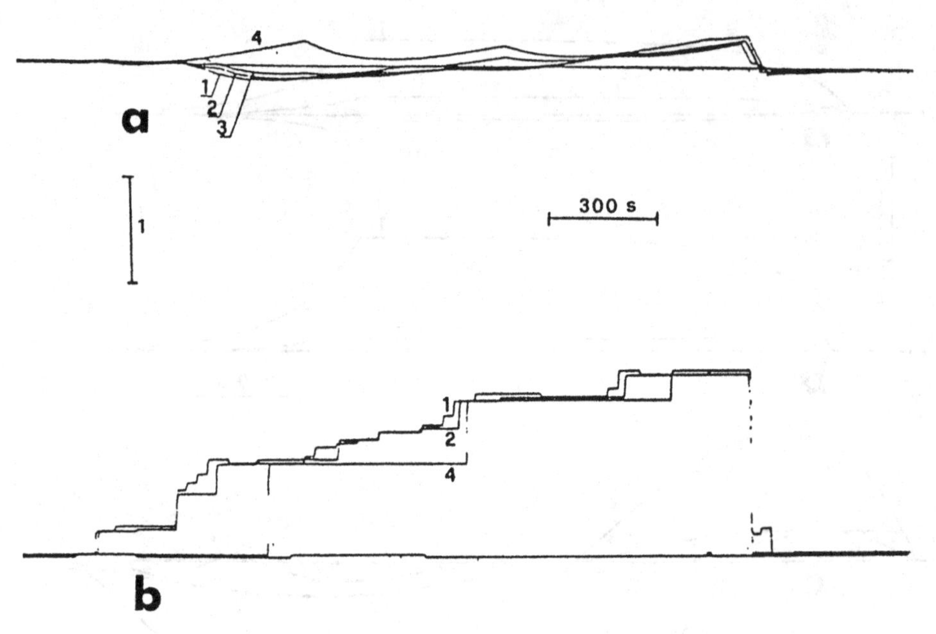

Figure 5. a) Process response for fuzzy feedforward control, saw-tooth disturbance and following sampling intervals: (1) T = 19.2 s, (2) T = 100 s, (3) T = 300 s and (4) T = 560 s
b) Control signals

The robustness of the fuzzy feedforward control to the variation of process fuzzy models parameters has been experimentally analyzed too. As an example Fig.6 shows process responses and a selection of corresponding control signals for the fuzzy feedforward control , saw tooth disturbance and various process fuzzy models. The values of the process fuzzy matrices $\underline{P}^*$ and $\underline{R}^*$ have been changed as follows:

1) Original form of $\underline{P}^*$ and $\underline{R}^*$ given with (11) and (12),

2) All values of $\underline{P}^*$ increased 50 %, $\underline{R}^*$ unchanged,

3) All values of $\underline{R}^*$ increased 50 %, $\underline{P}^*$ unchanged ,

4) All values of $\underline{P}^*$ and $\underline{R}^*$ increased 50 %,

5) All nonzero values of $\underline{P}^*$ and $\underline{R}^*$ changed to one. This means replacement of fuzzy matrices with Boolean matrices

$$\underline{P}^* = \begin{bmatrix} 0&0&0&0&0&0&0 \\ 1&0&0&0&0&0&0 \\ 1&1&1&0&0&0&0 \\ 0&0&1&1&1&1&0 \\ 0&0&0&0&1&1&1 \\ 0&0&0&0&0&0&1 \\ 0&0&0&0&0&0&0 \end{bmatrix} \quad \underline{R}^* = \begin{bmatrix} 1&0&0&0&0&0&0 \\ 1&1&0&0&0&0&0 \\ 0&0&1&0&0&0&0 \\ 0&0&0&1&0&0&0 \\ 0&0&0&0&1&0&0 \\ 0&0&0&0&0&1&1 \\ 0&0&0&0&0&0&1 \end{bmatrix} \tag{18}$$

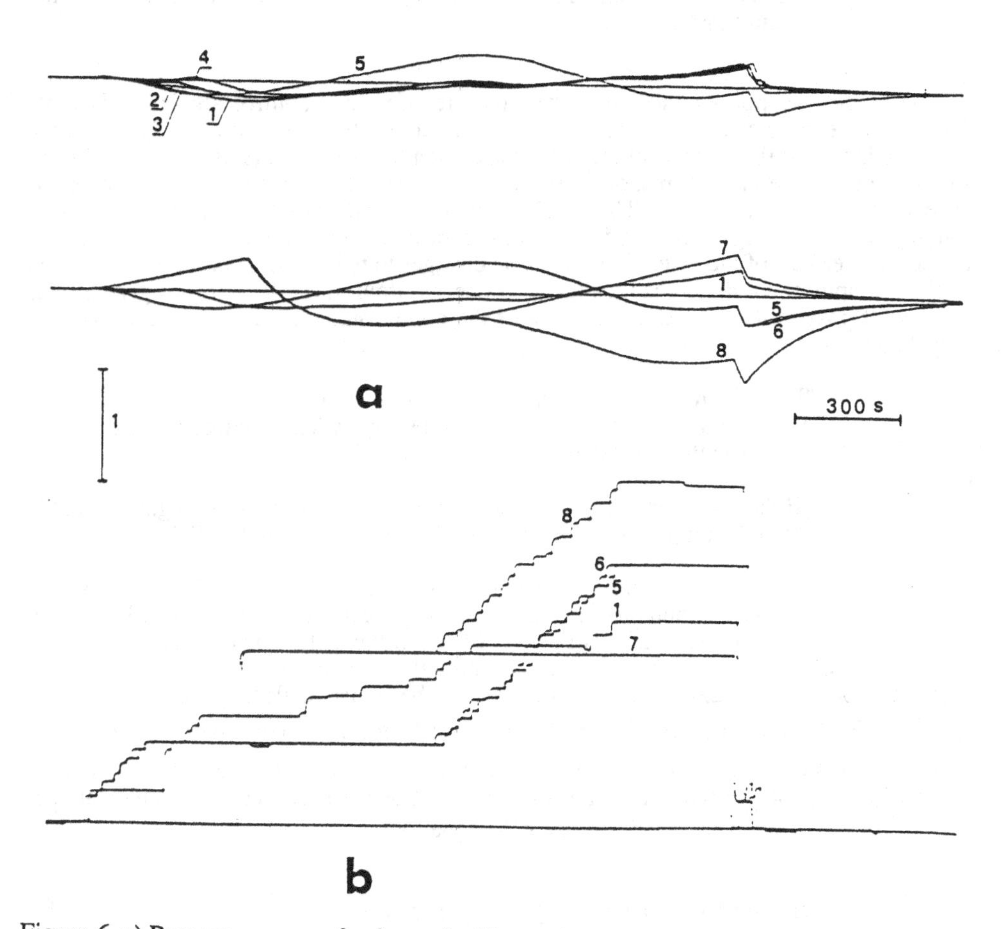

Figure 6. a) Process response for fuzzy feedforward control, saw-tooth disturbance and: (1) original fuzzy matrices $\underline{P}^*$ and $\underline{R}^*$, (2) $\underline{P}^*$ increased 50%, $\underline{R}^*$ unchanged, (3) $\underline{P}^*$ unchanged, $\underline{R}^*$ increased 50%, (4) $\underline{P}^*$ and $\underline{R}^*$ increased 50%, (5),(6) (7) and (8) $\underline{P}^*$ and $\underline{R}^*$ Boolean matrices (see text)
b) Selected control signals

6) All values of $\underline{P}^*$ and $\underline{R}^*$ bigger or equal to 0.1 replaced with 1 and values less then 0.1 replaced with 0,

7) All values of $\underline{P}^*$ and $\underline{R}^*$ bigger or equal to 0.4 replaced with 1 and all values less then 0.4 replaced with 0,and

8) In each row of $\underline{P}^*$ and $\underline{R}^*$ only the biggest value is replaced with 1 and all other values are set to 0. This means that in each row there is just one non zero element. Exception is the third row of matrix $\underline{P}^*$, because in its original form there are two values equal to 1. The matrix $\underline{R}^*$ is now the unity matrix.

The process responses show that the fuzzy feedforward control algorithm is quite flexible as regards changes to the process fuzzy model. For the cases 1) , 2) and 3) the response is almost the same as in original case. When the fuzzy model is replaced with the Boolean model the process response is not so good as before, but it is still satisfactory, especially for cases 5) and 6). The Boolean model only describes the process behavior roughly. In identifying such a model it is sufficient to evaluate and linguistically express the possible effects of the disturbance variable changes to the process output variables and of the manipulate variable changes to the process output variable, without giving their degrees of possibility. For example some of the conditional statements for the Boolean model are:

"**If** the increment of the disturbance input variable (Δd) is 'positive big', **then** the output increment (Δy) will be somewhere between 'positive small'' and 'positive medium' "

or

"**If** the increment of the manipulated input variable (Δu) is 'negative small' **then** the output increment (Δy) will be also 'negative small' "

As a result of further process model simplification (case 7. and particularly case 8.) the worse process response is obtained according to the process response obtained for basic Boolean model (case 5.), especially for higher values of the disturbance signal.

Influence of various inverse relational-relational composition operators used in the algorithm for solving fuzzy relational equations has been also analysed. Process responses shown in Fig.7. correspond to the saw tooth disturbances, original algorithm where α-composition is used as an inverse composition (eq.13) and modified algorithm where α-composition is replaced with max-min composition. For this second case the element u_j of the control vector $\underline{\Delta u}^*$ was calculated using the equation

$$u_j = \sup_l \{ \min[\sup_m (\min(p_{lm}, \Delta d_m)) , r_{lj}] \} \qquad (19)$$

When it was not possible to calculate u_j from the equation (13) or (19) the algorithm for approximate,numerical solution of fuzzy relational equations [6] was used.

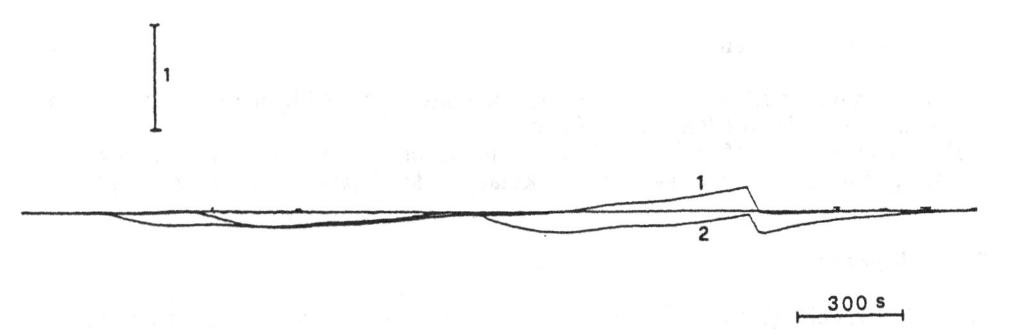

Figure 7. Process response for saw tooth disturbance and fuzzy feedforward control with (1) α-composition and (2) max-min composition used for solving fuzzy relational equations

In both cases process responses were quite satisfactory. Max-min composition gave even better results for higher values of disturbance signals, although according to the fuzzy mathematics α-composition must be used as an inverse composition for solving fuzzy relational equations with max-min composition [9].

Analyzing the experiment results we can conclude that in the same conditions, which means the same vague, non-accurate input-output data used in process model identification, the fuzzy feedforward approach gives better results then the conventional feedforward approach. Another advantage of the fuzzy feedforward control is controller robustness to the changing of the sampling interval , to the changing the process fuzzy model and to the choice of the inverse relational-relational composition operator used in algorithm for solving fuzzy relational equations.

6. Conclusion

The proposed and develop methods of fuzzy feedforward control are rather simple and effective, useful and applicable in all situations when only subjectively interpreted, inaccurate and linguistically expressed information about process behavior or disturbance variables values are known.

Experiments with a fuzzy feedforward controller shows that it gives better results then the conventional feedforward controller derived from the same data about the process behavior. Also the fuzzy feedforward controller is fairly robust to the changing of the sampling interval and the process fuzzy model parameter variations. Also the choice of the inverse relational-relational composition operator used in the algorithm for solving fuzzy relational equations is not so critical.

At the end it should be emphasis that the fuzzy feedforward control is not a substitute for the conventional feedforward control. It is only an alternative approach for invariant , disturbance decoupling control of ill-defined processes where conventional methods don't give the desired results.

Acknowledgment

The research presented in this paper has been partially supported by the Self-Management Community for Scientific Research of Croatia.

The authors thank to Mr.T.Kilic and Mrs.T.Stipica for their assistance with development and setup of the experimental system and specially in development of computer programs.

7. References

[1] Ulanov, G.M. (1964) *'Disturbance control and the principle of invariance'* in R.H.Mac (ed.),Progress in Control Engineering, A Heywood Book,London

[2] Stipanicev,D. (1987) *' Fuzzy Invariant Control of Complex Processes'*, Ph.D.Thesis, University of Zagreb, Zagreb

[3] Stipanicev,D. and Bozicevic,J. (1985) 'Fuzzy linguistic models for feedforward control', *Proc.of 13th IASTED Symp. on Modelling and Simulation*, Acta Press , Anahain , 211-214

[4] Stipanicev,D. and Bozicevic,J. (1986) 'Models of non-deterministic systems for application in feedforward control',*Proc.of 2nd European Simulation Congress*, Antwerpen, 240-246

[5] Pedrycz,W. (1984) 'Identification in fuzzy systems',*IEEE Trans.System,Man and Cybernetics*, Vol.14, No.2, 361-366

[6] Stipanicev,D. and Bozicevic,J. (1986) 'Fuzzy feedforward and composite control', *Trans. Inst. Measurement and Control*, Vol.8, No.2, 67-75

[7] Stipanicev,D., Bozicevic,J. and Mandic,I. (1987) 'Fuzzy feedforward-feedback computer control of processes with a factor of uncertainty', *Proc. of CEF87: The Use of Computer in Chemical Engineering*, Taormina, 669-679

[8] Malvache,D. and Willaes,N. (1981) 'The use of fuzzy sets for the treatment of fuzzy information by computer' , *Fuzzy Sets and Systems*, Vol.5, 323-327

[9] Sanchez,E. (1976) 'Resolution of composite fuzzy relational equations', *Inform. Control*, No.30, 38-48

THEORY AND DECISION LIBRARY

SERIES D: SYSTEM THEORY, KNOWLEDGE ENGINEERING AND PROBLEM SOLVING

Already published:

Topics in the General Theory of Structures
Edited by E. R. Caianiello and M. A. Aizerman
ISBN 90-277-2451-2

Nature, Cognition and System I
Edited by Marc E. Carvallo
ISBN 90-277-2740-6

Fuzzy Relation Equations and Their Applications to Knowledge Engineering
by Antonio Di Nola, Salvatore Sessa, Witold Pedrycz, and Elie Sanchez
ISBN 0-7923-0307-5

Fuzzy Sets in Information Retrieval and Cluster Analysis
by Sadaaki Miyamoto
ISBN 0-7923-0721-6

Progress in Fuzzy Sets and Systems
Edited by Wolfgang H. Janko, Marc Roubens, and H.-J. Zimmermann
ISBN 0-7923-0730-5